BIOTECHNOLOGY IN AGRICULTURE, INDUSTRY AND MEDICINE

DIATOMS

DIVERSITY AND DISTRIBUTION, ROLE IN BIOTECHNOLOGY AND ENVIRONMENTAL IMPACTS

BIOTECHNOLOGY IN AGRICULTURE, INDUSTRY AND MEDICINE

Additional books in this series can be found on Nova's website
under the Series tab.

Additional e-books in this series can be found on Nova's website
under the E-book tab.

DIATOMS

DIVERSITY AND DISTRIBUTION, ROLE IN BIOTECHNOLOGY AND ENVIRONMENTAL IMPACTS

FLAUBERT C. BOUR
EDITOR

NOVA BIOMEDICAL

New York

For permission to use material from this book please contact us:
Telephone 631-231-7269; Fax 631-231-8175
Web Site: http://www.novapublishers.com

NOTICE TO THE READER

The Publisher has taken reasonable care in the preparation of this book, but makes no expressed or implied warranty of any kind and assumes no responsibility for any errors or omissions. No liability is assumed for incidental or consequential damages in connection with or arising out of information contained in this book. The Publisher shall not be liable for any special, consequential, or exemplary damages resulting, in whole or in part, from the readers' use of, or reliance upon, this material. Any parts of this book based on government reports are so indicated and copyright is claimed for those parts to the extent applicable to compilations of such works.

Independent verification should be sought for any data, advice or recommendations contained in this book. In addition, no responsibility is assumed by the publisher for any injury and/or damage to persons or property arising from any methods, products, instructions, ideas or otherwise contained in this publication.

This publication is designed to provide accurate and authoritative information with regard to the subject matter covered herein. It is sold with the clear understanding that the Publisher is not engaged in rendering legal or any other professional services. If legal or any other expert assistance is required, the services of a competent person should be sought. FROM A DECLARATION OF PARTICIPANTS JOINTLY ADOPTED BY A COMMITTEE OF THE AMERICAN BAR ASSOCIATION AND A COMMITTEE OF PUBLISHERS.

Additional color graphics may be available in the e-book version of this book.

Library of Congress Cataloging-in-Publication Data

Library of Congress Control Number: 2013949905

ISBN: 978-1-62948-210-1

Published by Nova Science Publishers, Inc. † New York

Contents

Preface

In this book, the authors present current research in the study of the diversity and distribution, role in biotechnology and environmental impacts of diatoms. Topics discussed in this compilation include the determination of biogenic aluminum and rare earth element composition in diatom opal and its implication for marine chemistry of diatom frustule, an impure entity; diatom flora of fresh and brackish water bodies in the Far East of Russia; paleoceanogaphy since the warm pliocene epoch in the mid-latitudes of the Northwestern Pacific Ocean; the analysis of lake ecosystems transformation by taxonomic structure of diatom assemblages; and microphytobenthic community dynamics in the intertidal flats of the Nakdong River Estuary in South Korea.

Chapter I – A dissolution kinetics model of settling particles suggests that the siliceous matter in settling particles has a higher tendency to retain its original chemistry at a higher diatom flux. In this study, rare earth element (REE) composition of diatom opal was estimated by selecting less-contaminated and less-altered samples, based on their Al concentration and the dissolution kinetics, from settling particles collected in the Bering Sea, which has extremely high diatom productivity. The aluminum concentrations of a new set of settling particles were determined to evaluate terrigenous contamination and/or alteration of each sample. The Al concentration of siliceous fractions of the selected samples was lower than 0.21%, which is as small as that of cultivated diatoms. The concentration of aluminium in original diatom opal is not well understood. Hence, two assumptions were introduced: first, innate Al is assumed to be present (innate assumption), and second, innate Al concentration is assumed to be zero (terrigenous assumption). The REE concentration obtained in the innate assumption (high estimate) is higher than that of the terrigenous assumption (low estimate). The high estimate is chosen as the appropriate composition of original diatom opal for the following reasons: i) presence of REEs in diatom opal seems to be compatible with presence of Al, ii) the elemental mapping images of several diatom frustules exhibited a rather dispersed distribution of Al, iii) dissolution of diatom opal using the high estimate would release sufficient REEs to explain the observed concentration increase in deep water, iv) the oversupplied REEs in (iii), i.e., the difference between the dissolved REE composition of seawater reproduced by the dissolution of diatom opal with high estimate and that observed, maintains an almost identical partitioning pattern against the composition of seawater, which is similar to that ubiquitously seen between the acetic acid-soluble fraction of particles and seawater, v) a box model consideration using the high estimate gives a reasonable residence time of each REE in the water column, which is close to that obtained by independent methods, and vi) the REE incorporation theory reproduces a similar composition to the high estimate. The proposed concentration of REEs in diatom opal was lower than that in terrigenous matter (1/30–1/13), but was similar to chondritic levels. It exhibited marked enrichment in heavier rare earth elements relative to shale and in lighter rare earth elements relative to chondrite. It should be free from alteration as well as contamination with terrigenous matter. The diatomaceous REE data explain the nutrient-type vertical distribution of REEs in water columns. These data, in conjunction with fractionation due to carbonate/oxide scavenging, explain heavier REE-enrichment of deep water.

Chapter II – The results of diatom flora investigation from water bodies of the Sakhalin Island are presented herein. Sakhalin, the biggest island of Russia, is washed by the Sea of Okhotsk and the Sea of Japan. It is separated by the Strait of Tartary from continental Asia (Russia) and by the La Pérouse Strait from the Hokkaido Island (Japan). Water bodies of Sakhalin (rivers, springs, fresh and brackish water lakes, hot spring) are represented by variety of types and sizes, and differ from each other by water temperature, pH, mineralization and trophic level. As a result, the diversity of diatom algae on Sakhalin is quite high. Similarities and differences of dominant diatom taxa and community composition between each type of water bodies are investigated and presented in this chapter. Over of 500 species, varieties and forms of diatoms from three classes (Coscinodiscophyceae, Fragilariophyceae, and Bacillariophyceae) are recorded for the island. New records of diatoms for water bodies of the Sakhalin Island are reported. Ecological and geographical characteristic (relation to habitat, salinity, pH, saprobity, and geographical distribution) of the Sakhalin diatom flora are discussed.

Chapter III – Diatom-derived sea surface temperatures (Td'-SST) ($^{\circ}$C) after 5.3 Ma gradually decreased with pronounced warm peaks at 4.5-4.3, 3.7-3.3, and 3.1 Ma at DSDP Sites 578-580 and Site 436 in the middle latitudes of the northwestern Pacific Ocean. At northern site 436, estimated paleotemperatures based on the comparison between Twt and Td'-SSTs ($^{\circ}$C) have large differences from present-day mean annual temperatures ranging from 22 °C to 10°C. The values at southern Site 578, in contrast, show small fluctuations with the largest difference of 4.5°C. Fluctuations of Td'-SSTs ($^{\circ}$C) and diatom assemblages suggest influences of Earth's orbital forcing cycles.

Diatom abundances (10^{7} valves/g) and Td'-SSTs ($^{\circ}$C) decrease after 4.8 Ma at Site 579 in the modern Mixed Water Region. The relative abundances of subtropical-warm diatom species *Thalassionema nitzschioides* s.l. predominate occupying 40-60 % of the diatom assemblages in the early Pliocene, and 15-45 % in the late Pliocene at Site 579. And remarkable decrease occurs at 2 Ma. The relative abundances of sub-littoral upwelling diatom *Chaetoceros* spores increase by 5-20 % over after 2 Ma. During the Pliocene epoch, cold upwelling front may be absent in the subtropical-warm water region because latitudinal thermal gradient between the equator and the subtropical/ middle latitudes was very weak during the warm Pliocene epoch.

Chapter IV – The biomass dynamics of microphytobenthos (MPB) in the intertidal flats of the Nakdong River estuary (South Korea) were studied from August 2006 to August 2008. On spatial and temporal distribution, the MPB exhibited evident seasonal variation among different sampling sites. Muddier sediments possessed higher biomass which had several peaks in spring, summer or winter, while in the sandiest sediment, MPB biomass was lowest and had only one peak in summer. Multivariate correlation analysis revealed that the biomass was positively correlated with mud (< 63 µm) and very fine sand (63–125 µm) composition, and negatively related to fine sand (125–250 µm) and medium sand (≥ 250 µm) composition, but not statistically related to dissolved inorganic nutrient concentration (NH_4^+, NO_2^- + NO_3^-, PO_4^{3-}, and $Si(OH)_4$), salinity and light intensity. On vertical distribution, MPB biomass declined exponentially with depth, and also related to sediment type. In muddier sediments, the exponential curve slope was larger than sandier sediments, implying lower resuspending and burying speed and higher stability of sediments. Diel vertical migrations presented a certain degree of rhythm following tidal and light cycles, viz. surfacing in low-tide emersion at daytime, migrating down at nighttime and/or high-tide submersion. However,

MPB migratory rhythm also primarily determined by their biomass, which dramatically weakened the rhythm when visibly thick biofilms formed in winter: MPB cells remained at the surface, even at night and/or during high tide. Generally, our study revealed MPB community dynamics in intertidal flats as results from interactions between biotic and environmental factors.

In: Diatoms
Editor: Flaubert C. Bour

ISBN: 978-1-62948-210-1
© 2013 Nova Science Publishers, Inc.

Chapter I

A Diatom Frustule is an Impure Entity: Determination of Biogenic Aluminum and Rare Earth Element Composition in Diatom Opal and Its Implication for Marine Chemistry

Tasuku Akagi[*1], *Mariko Emoto*[1], *Rie Takada*[1] *and Kozo Takahashi*[1,2]

[1]Department of Earth and Planetary Sciences, Faculty of Sciences, Kyushu University, Higashi-ku, Fukuoka, Japan
[2]Hokusei Gakuen University, Atsubetsu-ku, Sapporo, Japan

Abstract

A dissolution kinetics model of settling particles suggests that the siliceous matter in settling particles has a higher tendency to retain its original chemistry at a higher diatom flux. In this study, rare earth element (REE) composition of diatom opal was estimated by selecting less-contaminated and less-altered samples, based on their Al

[*] Corresponding author: Tasuku Akagi, E-mail address: akagi@geo.kyushu-u.ac.jp.

concentration and the dissolution kinetics, from settling particles collected in the Bering Sea, which has extremely high diatom productivity. The aluminum concentrations of a new set of settling particles were determined to evaluate terrigenous contamination and/or alteration of each sample. The Al concentration of siliceous fractions of the selected samples was lower than 0.21%, which is as small as that of cultivated diatoms. The concentration of aluminium in original diatom opal is not well understood. Hence, two assumptions were introduced: first, innate Al is assumed to be present (innate assumption), and second, innate Al concentration is assumed to be zero (terrigenous assumption). The REE concentration obtained in the innate assumption (high estimate) is higher than that of the terrigenous assumption (low estimate). The high estimate is chosen as the appropriate composition of original diatom opal for the following reasons: i) presence of REEs in diatom opal seems to be compatible with presence of Al, ii) the elemental mapping images of several diatom frustules exhibited a rather dispersed distribution of Al, iii) dissolution of diatom opal using the high estimate would release sufficient REEs to explain the observed concentration increase in deep water, iv) the oversupplied REEs in (iii), i.e., the difference between the dissolved REE composition of seawater reproduced by the dissolution of diatom opal with high estimate and that observed, maintains an almost identical partitioning pattern against the composition of seawater, which is similar to that ubiquitously seen between the acetic acid-soluble fraction of particles and seawater, v) a box model consideration using the high estimate gives a reasonable residence time of each REE in the water column, which is close to that obtained by independent methods, and vi) the REE incorporation theory reproduces a similar composition to the high estimate. The proposed concentration of REEs in diatom opal was lower than that in terrigenous matter (1/30–1/13), but was similar to chondritic levels. It exhibited marked enrichment in heavier rare earth elements relative to shale and in lighter rare earth elements relative to chondrite. It should be free from alteration as well as contamination with terrigenous matter. The diatomaceous REE data explain the nutrient-type vertical distribution of REEs in water columns. These data, in conjunction with fractionation due to carbonate/oxide scavenging, explain heavier REE-enrichment of deep water.

1. Introduction

Diatoms contribute to more than half of the primary production of the oceans and it is well established that the formation and dissolution of diatom opal governs the distribution of dissolved silica in ocean columns (Nelson et al., 1995). Owing to the physical and chemical difficulty in isolating diatom opal from clays (Shemesh et al., 1988; Beck et al., 2002), however, it is difficult to clarify the trace composition of opal and thus to understand how diatoms contribute to the ocean circulation of trace elements. To date, no direct determination of rare earth elements (REEs) in diatoms has been made, and the role of diatoms has not been considered in the circulation of REEs (Sholkovitz et al., 1994; Oka et al., 2009; Siddall et al., 2008; Arsouze et al., 2009; Tachikawa et al., 2003).

We have reported that diatom opal is subject to non-linear dissolution kinetics, in which the dissolution rate of elements increases exponentially with decreasing production rate (Akagi et al., 2011). The kinetic theory is consistent with the observation that opal sedimentation occurs only in areas of high diatom productivity (Nelson et al., 1995). In such highly productive areas, diatom opal may undergo very little dissolution because of its high settling velocity and small specific surface areas, both of which are due to efficient aggregation (Akagi et al., 2011). Therefore, the original REE composition of diatom opal can be estimated using appropriate samples selected on the basis of dissolution kinetics. The North Pacific Ocean has very high silica concentration in deep water (Tréguer et al. 1995). In particular, the Bering Sea is known to contain the highest silica concentration in the world (Edmond et al., 1979; Tsunogai et al., 1979), and thus this area is extremely productive (Takahashi et al., 2002; Sambrotto et al., 1984). Hence, in this study, dissolution kinetics was applied to sediment trap samples in the Bering Sea and the data for REE compositions in diatom opal will be discussed after considering the presence of Al in diatom opal.

2. Theory and Method

Settling particles consist of opal, aluminosilicates, carbonates, oxides and organic matter. We first separated opal and aluminosilicate from the others by treating settling particles with 40% acetic acid and named the separates "siliceous fractions". The composition of the siliceous fraction can be modified from the original composition of diatom opal by contamination from clay matter as well as dissolution.

The contribution of dissolution may be grasped using the dissolution kinetics of opal particles. Dissolution kinetics leads to the following equation (Akagi et al., 2011).

$$\frac{C_j}{C_i} \approx \frac{C_j{}^0}{C_i{}^0}\left[1+\frac{KD}{P^2}\right]$$

$$\tag{1}$$

where C_i, K, P and D are the concentration of element i in opal particles, proportionality constant, particle density (or diatom productivity), and the distance that particles travel, respectively. The superscript "0" represents the original concentration. Considering the case that where i is Si, eq. (1) indicates that the original concentration of element j in opal particles can be estimated from samples with high-enough P values.

The North Pacific Ocean has a very high deep-water silica concentration (Tréguer et al. 1995). In particular, the Bering Sea is known to contain the highest silica concentration in the world (Edmond et al., 1979; Tsunogai et al., 1979), and thus this area is extremely productive (Takahashi et al., 2002; Sambrotto et al., 1984). The Bering Sea is therefore a prime area for study with samples at high-enough P values being available. We used the reported REE data of siliceous fraction of settling particles trapped at two stations in the North Pacific Ocean including the Bering Sea (Akagi et al., 2011): Station SA, centrally located at a subarctic pelagic station (49°N, 174°W; water depth 5406 m), and Station AB, located at a marginal sea station (53.5°N, 177°W; water depth 3788 m). The water of Station AB is more productive than that of Station SA and the opal flux at Station AB is about twice that of Station SA (Takahashi et al., 2002).

Aluminum is one of the most representative elements of terrigeouns matter, although it is not sure that Al in diatom opal is always due to terrigenous contamination. In this paper to determine Al we used a new set of 1/1024 aliquots of the same batch of slurries as used to determine REEs in the early work (Akagi et al., 2011). Briefly, the sample slurry was treated successively with 10 ml of ethanol and 10 ml of 40% acetic acid solution. The residue was washed with Mili-Q water and transferred to a Teflon vessel and was digested with 1ml of 6 M HNO_3, 1 ml of 40% HF and 1 ml of 60% $HClO_4$ acids. After evaporating to dryness, the HNO_3 solution was added to prepare solutions for ICP-AES measurement of Al and ICP-MS measurement of La. Lanthanum was determined to correct the heterogeneity of the aliquot by normalizing the concentration of La in the 1/1024 division to that in the previous 1/256 division. We assumed that the composition of the siliceous fraction is $SiO_2+0.4H_2O$ and that Si content of siliceous matter is constantly 41.7%.

For SEM (scanning electron microprobe) elemental mapping, 1/4096 or 1/16384 aliquots (approx. 40 ml with dilute formaldehyde solution) of other sediment trap samples (AB18#1, #6 and #16) were used. They were collected from Station AB in the Bering Sea in 2006 and 2007. The AB18#1, #6 and #16 samples were collected in highly, moderately and less productive months (330, 150 and 8.9 mg opal $m^{-2}d^{-1}$), respectively. The three sample slurries were first heated with 10 ml of 30% $HClO_4$ and 10 ml of 1 M HCl solutions and the heavily-aggregated residue was collected and dispersed in a Calgon® solution and filtered to collect on a membrane filter using pure water. The diatom frustule samples on the dried filter were picked using a hair under a microscope and transferred to resin on a glass plate. The resin was consolidated at 40 °C and polished carefully to expose the diatom opal sections using diamond paste. The carbon-coated samples were analysed with EPMA equipped with a field-emission electron gun (JEOL JXA-8530F) at an irradiation current of 1.005×10^{-8} A and accelerating voltage from 15 to 20 kV. $K\alpha$ lines were used for detection of Si and Al.

3. Results and Discussion

3.1. Al Concentration and Selection of Less Contaminated Samples

The Al concentrations in the siliceous fraction of the settling particles collected from the North Pacific Ocean are listed in Table 1. The concentrations of Al were highly variable and ranged from 700 to 8000 µg/g. There was no significant difference in Al concentration between the two stations. In more productive months the concentration of Al tended to be smaller than in less productive months. The relationship between ΣREE and Al concentration of the settling particles are shown in Fig. 1. Generally the ΣREE values are greater with higher Al concentrations and data plots approach the linear relationship of loess (e.g. Ding et al., 2001). It is interesting that the ΣREE values distribute around 5 µg/g and only slightly increase with the increasing Al concentration when Al concentrations < 5000 µg/g. Assuming that loess is responsible for the Al in the siliceous fractions, the ΣREE values were corrected for using the Al concentrationin the fractions and a typical loess composition (e.g. Ding et al., 2001) (Al-corrected ΣREE). Because the most of plots distributed below the loess line, the Al-corrected ΣREE values are smaller than zero. Since the influence of terrigenous contamination of REEs cannot be ruled out in any of the samples, we selected the three samples from Station AB shown in the dotted square in Fig. 1: AB6#1, AB6#11 and AB6#13. The three samples are unique because Al concentration is low enough and the Al-correction represents only about half the REE concentration of the siliceous fractions. It will be shown in the next section that the three samples are also least altered by dissolution. The Al concentration of the selected samples forms the lowest concentration group. Dixit et al. (2001) reported Al/Si (mol/mol) ratio of 0.0075 ±0.0048 for natural trapped diatom samples using SEM. Beck et al. (2002) reported low ratios from 7×10^{-5} to 0.007 for cultured samples and 0.008 for one natural sample using PIXE or AAS. The selected three samples have Al/Si of 0.0042 ±0.0014 (mol/mol).

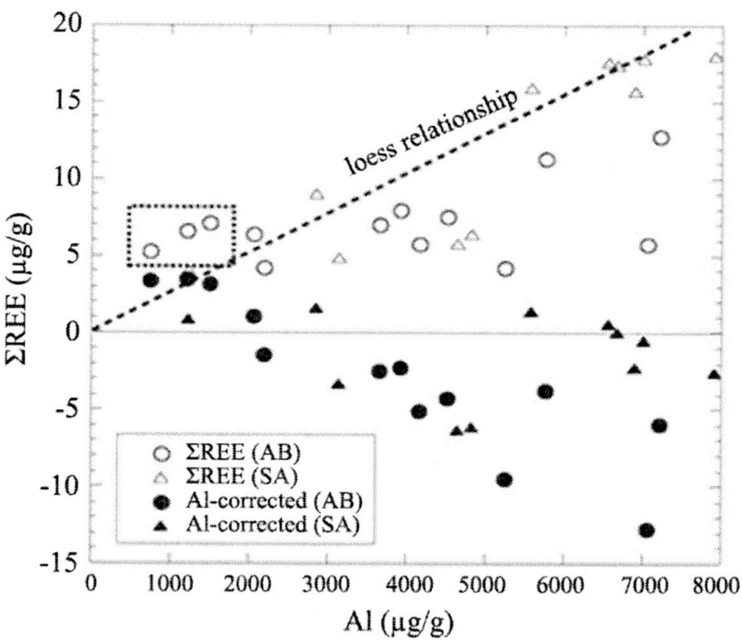

Figure 1. Relationship between ΣREE and Al concentration in siliceous fractions of the settling particles from Stations AB (open circles) and SA (open triangles) in the North Pacific Ocean. The data shown in closed marks are obtained after the REE data are subtracted by the assumed terrigenous contamination based on Al concentration. The broken line represents the relationship for loess input (Ding et al., 2001), a representative of terrigenous matter. The three samples from Station AB in the dotted square are selected as appropriate samples least contaminated with terrigenous matter.

Table 1. Concentration of aluminum in settling particles along with ΣREE and opal flux data

Station AB	6#1	6#2	6#3	6#4	6#5	6#6	6#7	6#8	6#9	6#10	6#11	6#12	6#13
Mid Date	15 Aug.	2 Sep.	22 Sep.	12 Oct.	9 Nov.	25 Dec.	19 Feb.	29 Mar.	18 Apr.	8 May	28 May	17 Jun.	13 Jul.
Al concn. (µg/g)	724	5250	4150	4520	3550	5750	7200	7050	1500	2180	1200	3920	2150
ΣREE (µg/g)[a]	5.29	4.26	5.79	7.61	7.03	11.3	12.8	5.75	7.13	4.23	6.62	7.95	6.39
Opal flux (mg m^{-2} d^{-1})[a]	221.7	89.5	79.0	97.0	95.3	61.3	48.4	106.8	605.2	805.0	372.7	260.7	224.2
Station SA	7#1	7#2	7#3	7#4	7#5	7#6	7#7	7#8	7#9	7#10	7#11	7#12	7#13
Mid Date	26 Aug.	22 Sep.	12 Oct.	3 Nov.	19 Dec.	19 Feb.	29 Mar.	18 Apr.	8 May	28 May	17 Jun.	7 Jul.	25 Jul.
Al concn. (µg/g)	7920	6680	5570	6890	6550	7000	4810	4640	3130	2110	1220	2770	2840
ΣREE (µg/g)[a]	18.1	17.4	15.9	15.8	17.6	17.8	6.46	5.79	4.83		4.03		8.98
Opal flux (mg m^{-2} d^{-1})[a]	100.1	49.6	61.6	57.2	22.9	22.8	33.5	29.9	54.4	39.4	141.9	73.5	36.2

a) Data from Akagi et al. (2011).

Table 2. Estimation of REE composition (μg/g) of diatom opal

	La	Ce	Pr	Nd	Sm	Eu	Gd	Tb	Dy	Ho	Er	Tm	Yb	Lu	Al
Siliceous fraction of settling particles															
AB6#1[a]	0.912	2.03	0.245	0.956	0.239	0.0505	0.231	0.0329	0.205	0.0436	0.136	0.0230	0.157	0.0286	700
Al corrected	0.524	1.26	0.155	0.619	0.170	0.0362	0.167	0.0235	0.149	0.0321	0.103	0.0182	0.124	0.0238	
AB6#9[a]	1.21	2.70	0.336	1.29	0.283	0.0702	0.291	0.0497	0.314	0.0680	0.223	0.0316	0.229	0.0377	1500
Al corrected	0.40	1.10	0.151	0.594	0.139	0.0406	0.156	0.0301	0.198	0.0443	0.154	0.0215	0.162	0.0279	
AB6#10[a]	0.736	1.72	0.193	0.730	0.158	0.0402	0.160	0.0250	0.161	0.0353	0.114	0.0164	0.118	0.0185	2100
Al corrected	-0.33	-0.42	-0.053	-0.20	-0.033	0.0007	-0.017	-0.0011	0.006	0.0036	0.022	0.0030	0.028	0.0053	
AB6#11[a]	1.13	2.59	0.314	1.17	0.258	0.0651	0.271	0.0426	0.273	0.0580	0.192	0.0272	0.196	0.0311	1200
Al corrected	0.488	1.30	0.165	0.613	0.142	0.0414	0.164	0.0269	0.180	0.0390	0.136	0.0191	0.142	0.0232	
AB6#13[a]	1.13	2.65	0.296	1.06	0.226	0.0599	0.245	0.0373	0.238	0.0506	0.164	0.0251	0.183	0.0296	2000
Al corrected	0.06	0.51	0.049	0.13	0.034	0.0204	0.068	0.0111	0.083	0.0189	0.072	0.0118	0.093	0.0164	
Average diatom opal[b]															
High estimate (n=3)[c]	1.08±15	2.4±4	0.30±5	1.13±17	0.26±2	0.061±1	0.26±3	0.041±8	0.26±6	0.056±12	0.18±4	0.027±4	0.19±4	0.032±5	1000±400
(n=5)[d]	1.02±19	2.3±4	0.28±6	1.04±21	0.23±5	0.057±12	0.24±6	0.038±9	0.24±6	0.051±13	0.17±4	0.025±6	0.18±4	0.029±6	1600±600
Low estimate (n=3)[c]	0.47±6	1.2±1	0.157±7	0.61±1	0.15±2	0.039±3	0.16±5	0.027±3	0.17±2	0.038±7	0.13±3	0.020±2	0.14±2	0.025±3	0 (-)
Average loess[e]	36±1	71±3	8.2±3	31±1	6.4±4	1.32±4	5.9±0.3	0.87±6	5.2±0.4	1.1±0.1	3.1±0.3	0.45±4	3.0±3	0.44±4	6670±370

a) The original REE data of the five siliceous fractions are from ref.1. The total mass fluxes of the five samples are greater than 200 mgm^{-2}d^{-1} (Akagi et al., 2011).

b) The average data are the geometric mean of the three or five data ($\pm$ one σ). In the high estimate, uncorrected values are adopted assuming that all the detected Al is innate (innate assumption) and in the low estimate the values are corrected assuming that the Al is from terrigenous contamination (terrigenous estimate).

c) The data of the three samples with the three lowest Al concentrations (AB6#1, AB6#9, and AB6#11) were used.

d) The data of the five samples along the asymptotic lines in Fig. 2 (AB6#1, AB6#9, AB6#10, AB6#11 and AB6#13) were used.

e) Average loess data ($\pm$ one σ) are from Ding et al. (2001).

3.2. Selection of Less Altered Samples

In a previous study, five siliceous fractions trapped during the highly productive months (AB6#1, AB6#9, AB6#10, AB6#11, AB6#13; total mass flux > 300 mg m^{-2}day^{-1}) of the 24 samples showed consistent features in rare earth element composition, such as Ce anomalies, Eu anomalies, ΣREE and the slopes of REE patterns. They form an asymptote in the discussion based on the kinetic dissolution model (Akagi et al., 2011). The example of Ce anomaly, aberration of Ce from neighbouring La and Pr in a REE abundance pattern, is shown in Fig. 2 (Akagi et al., 2011). The five samples showed asymptotic values also for Eu anomaly, ΣREE and HREE index (slope of REE patterns). Incidentally, the other samples trapped mostly during less productive months showed a marked deviation from this composition. The five productive samples are considered to have undergone very little dissolution and/or alteration and thus were selected. All five samples were from Station AB in the Bering Sea, and none of the samples from the less-productive Station SA, located in the south of the Aleutian Islands, were selected.

The REE and Al compositions of the five samples are listed in Table 2.

3.3. High and Low Estimates of Diatom Opal

Settling particles generally comprise some terrigenous matter, in which REEs may be present at a higher concentration than in diatom opal. Hence, the contribution of such terrigenuous contaminants must be corrected to understand/estimate the composition of diatom opal. However, it is difficult to estimate the extent of terrigenous contamination because some elements contained in terrigenous matter may be innate in diatom opal (Beck et al., 2002). In light of such ambiguity regarding the composition of original diatom opal, we introduce two alternative assumptions to estimate the REE composition of original diatoms. The three samples we mentioned above had Al concentration less than 0.15%, which is as small as the concentration detected in opal structures in cultured and natural diatoms (Beck et al., 2002), and the rest of the 24 samples had Al concentrations ranging from 0.2% to 0.8%. Firstly, Al was assumed to be innate in the siliceous samples (*innate assumption*), and secondly, it was assumed to be terrigenous contamination (*terrigenous assumption*). The compositions of the three or five siliceous fractions were geometrically averaged to obtain the uncorrected estimate composition (Table 2), which corresponds to the high estimate for the innate assumption. The high estimate showed a HREE enrichment in the absence of a Ce anomaly (Fig. 3). The typical uncertainty of the estimation is about 20–30% from the variation of the data.

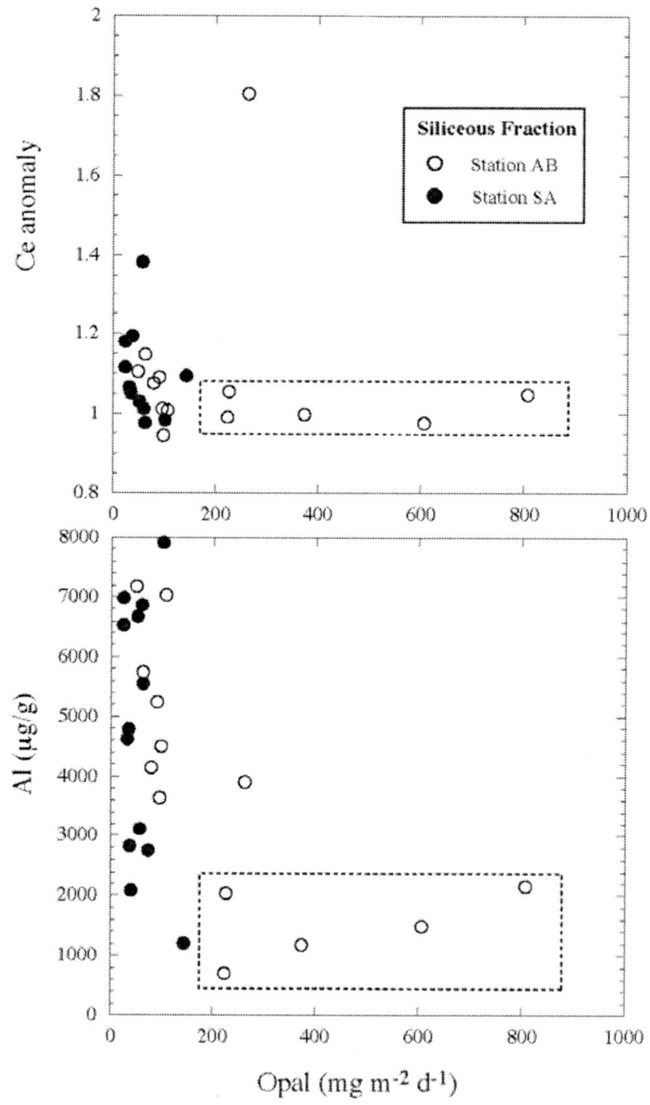

Figure 2. An example of the hyperbolic relationship observed for REE pattern (Ce anomaly) (a) and Al concentration (b) of siliceous fraction of settling particles with opal flux. Ce anomaly data are cited from Akagi et al., 2011. Five samples in the broken square were selected as the representatives of less-altered original diatom opal.

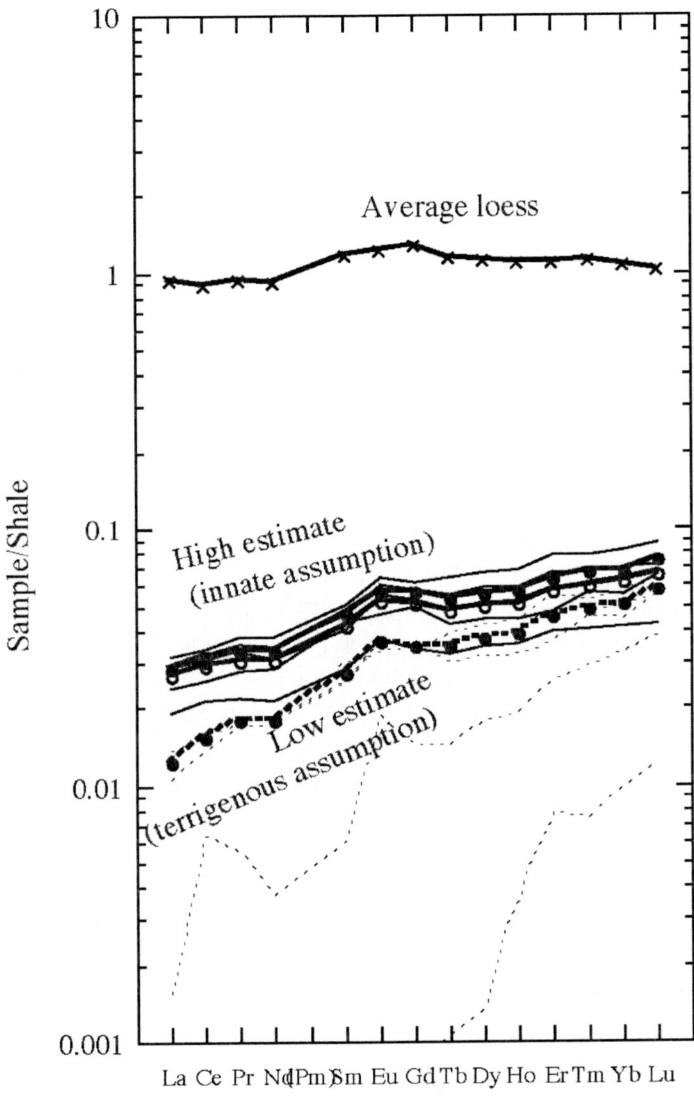

Figure 3. Shale-normalized abundance patterns of the estimated REE compositions of diatom opal. The high estimate assumes that Al is innate. Data for the open and closed circles are estimated from the five and three selected samples in Table 2, respectively. The low estimate assumes that Al is present as terrigenous contamination and was obtained by using the data of the three selected samples. The average loess pattern (Ding et al., 2001) from the Loess Plateau, China, used in the terrigenous correction is also shown. Normalization values are obtained from McLennan (1989).

The terrigenous assumption gives a low estimate. The estimate depends on the what kind of terrigenous matter is chosen for the estimation. The Bering Sea is on the path of airborne particle transport from China (Ginoux et al., 2001), and these particles were also observed on a Canadian mountain (Zdanowicz et al., 2006). Identical REE compositions were observed in loess samples from the Loess Plateau (36°56′N) in northern China and the airborne particles collected at the St. Elias Mountain (64°34′N) in Canada. The average composition of the loess samples (Ding et al., 2001) from the Loess Plateau was chosen as a representative of the terrigenous matter supplied to the sampling station in the Bering Sea (53°30′N) (Table 2). Aleutian rock is also a possible source of terrigenous matter in diatom opal. The REEs/Al ratio of the Aleutian rocks is smaller than that of loess (Ding et al., 2001; Kelemen et al., 2003). We adopted the loess as a representative of terrigenous matter to start with, because the correction for the terrigenous contamination using loess would provide the lowest possible REE composition.

When Al in the siliceous fraction was regarded as the terrigenous contamination (terrigenous assumption), the contribution of terrigenous matter was significant but not so high as to overwhelm the diatom contribution in case of the three samples with smaller Al concentrations. The terrigenous matter contribution in the terrigenous assumption was approximately 40%–60% at lighter REEs (LREEs) and 20%–30% at HREEs (Table 2). After correcting for the terrigenous contribution, the three samples had similar concentrations (Table 2; Fig. 3). The geometric mean of the Al-corrected compositions of the three samples was adopted as the REE composition of Al-corrected average diatom opal (Table 2). If an uncertainty of 10% is introduced in the composition of the terrigenous matter, which is sufficient to cover most of the variation of loess (Ding et al., 2001; Jahn et al., 2001), and the variation of the three data in Table 2 is considered, the estimation error is as much as 20% at LREEs and 10%–30% at medium REEs (MREEs) and HREEs. Al-corrected diatom opal is characterized with a marked HREE enrichment compared with shale (McLennan, 1989) (Fig. 2). Once again the low estimate corresponds to the lowest possible estimate, because of the choice of loess as a terrigenous contaminant of relatively high REEs/Al values.

In both assumptions, the estimated concentration level of REEs in diatom opal is lower than that in terrigenous matter (Ding et al., 2001; Jahn et al., 2001; McLennan, 1989) (1/30–1/80 at LREEs and 1/13–1/20 at HREEs). The true REE composition is very likely to fall somewhere between the two estimates. We consider that the high estimate is closer to the true value for the following reasons. 1) Since the concentration of REEs in diatom opal is not zero even in the terrigenous assumption, we have no logical reason for excluding Al, a trivalent ion like REEs, from opal. 2) If the diatom bloom occurs independently of terrigenous fallout (Takahashi et al., 2002), it is likely that the asymptote obtained by the kinetic theory may give the composition that already excludes the terrigenous contamination. 3) The dispersive nature of Al in opal does not seem to be due to attachment of clay. 4) The higher estimate can neatly explain the reported vertical profiles of dissolved REEs by considering scavenging by a carbonate phase. 5) The mean residence times of REEs obtained using the high estimate are more consistent with the reported values than if the low estimate is used. 6) The diatom incorporation theory developed based on selective intake of REE-silicic acid complexes in seawater predicts values close to the higher estimate. The last four points will be discussed in the following.

3.4. Distribution of Aluminium in Diatom Opal

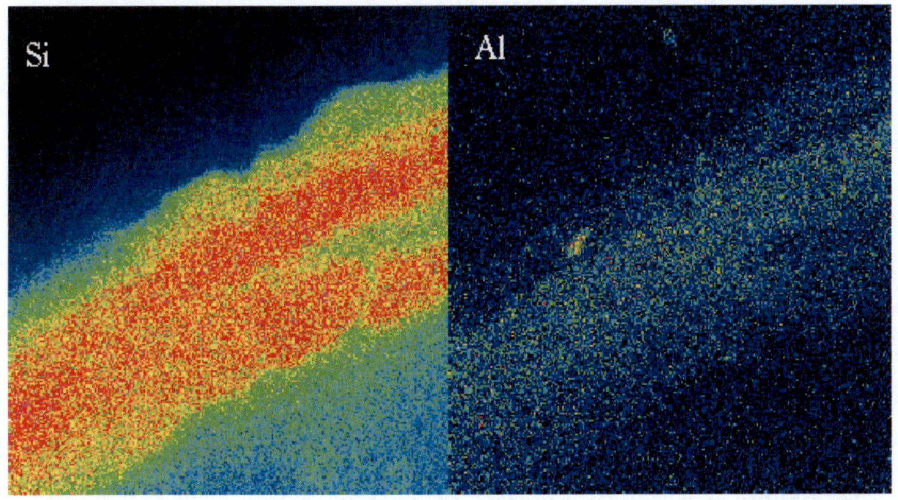

Figure 4. SEM images of Si and Al distributions of diatom frustules in a sediment trap sample (AB18#6) collected from the Bering Sea.

An example of SEM (scanning electron microprobe) images of aluminium and silicon in a frustule are shown in Fig. 4. The sample shown is a fragment of *Thalassiosira* sp., one of common diatoms in the Bering Sea. Significant variation in the intensity was observed among frustules from a sample, indicating compositional heterogeneity of each specimen. The highest signal of Al was detected from two frustules of the AB18#6 sample. The example shown in Fig. 4 is one of the two. The intensity of aluminium is homogeneously distributed, wherever the intensity of Si is strong. If it were distributed locally together with other cations, this would imply Al from clay attachment. Other specimens with less intensive Al signals also showed a similar dispersed distribution of Al. The dispersed distribution supports the innate assumption.

3.5. Contribution of Diatom Opal Dissolution to Vertical Distribution

Opal has been established as a Si carrier in seawater columns (Nelson et al., 1995). Diatoms are the major opal-bearing plankton, which forms about 50% of the primary production in the oceans (de la Rocha, 2006). Most of the REEs in diatom opal should be released into the water column during diatom dissolution. The release of REEs from opal dissolution (ΔREE_{opal}) can be estimated from the concomitant increase in Si concentration in the water column (ΔSi), i.e. ,

$$\Delta REE_{opal} = \left(\frac{REE}{Si}\right)_{opal} \times \Delta Si. \tag{2}$$

In Fig. 5 the ratio of the observed ΔREE (maximum REE concentration – surface REE concentration) to ΔREE_{opal} is plotted for each REE. It was found that either the high or low estimate of diatom opal, given in Table 1, can account for most of the increase of REEs in seawater columns in the North Pacific Ocean (Fig. 5). The concentration increase of LREEs and HREEs in the columns of the North Pacific Ocean (Alibo and Nozaki, 1999; Piepgras and Jacobsen, 1992, Zhang and Nozaki, 1996; DeBaar et al., 1985; Nozaki et al., 1999; Shimizu et al., 1994) is comparable to (low estimate; figure not shown) or more than half (high estimate; Fig. 5) the concentration increase that can be attributed to diatom dissolution, although larger variance is observed in the LREE region than in the HREE region. This indicates that the REEs in deep water are supplied primarily by diatom opal dissolution. The concentration increase of MREEs (Nd, Sm and Eu) is significantly lower than what can be explained by diatom dissolution. The low concentration may be either due to MREE removal from the water column or the difference in REE compositions between the Bering Sea and the North Pacific Ocean, or a combination of both. As discussed next, scavenging of REEs by carbonate/oxide/organic matter is likely to be responsible for the low concentration of MREEs.

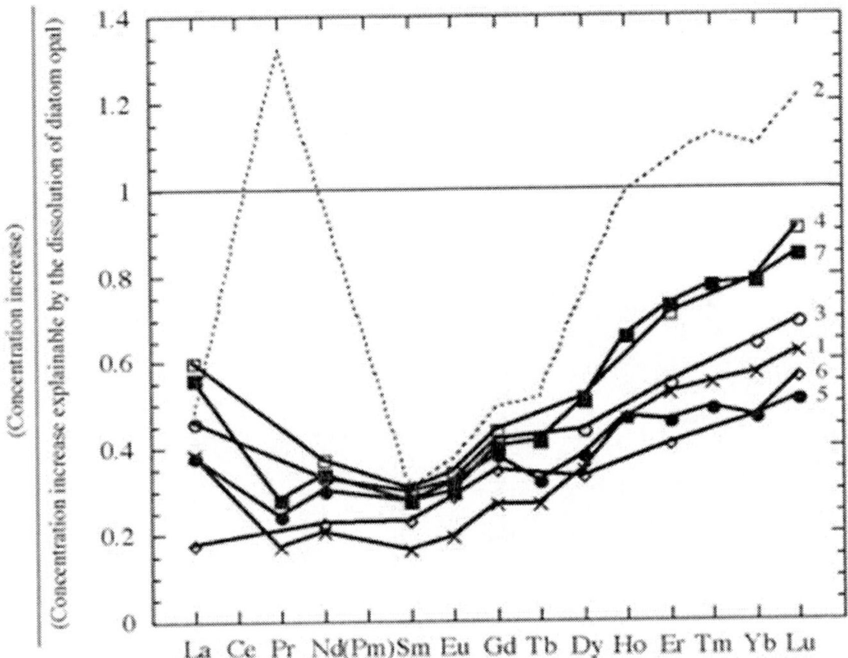

Figure 5. Ratio of the observed concentration increase in the North Pacific Ocean to that explained by diatom dissolution. Increase of dissolved Si is assumed to be due to diatom dissolution. Depth profiles, in which data for depths less than 100 m and more than 2900 m are given, are cited. One of the data (shown in broken lines) were by neutron activation analysis, whose precision may be worse than that of the other methods. Pacific Ocean data are from 1. Albo and Nozaki (1999) (34°41′N, 139°54′E), 2 DeBaar et al. (1985) (18°N, 108°W), 3-4 Piepgras and Jacobsen (1992) (47°00.0′N, 161°08.2′E; 24°17.2′N, 150°28.2′N), 5 Nozaki et al. (1999) (8°50′N, 121°48′E) 6 Shimizu et al. (1994) (44°40′N, 177°00′W), and 7 Zhang and Nozaki (1996) (34°30.32′N, 140°30.80′E). The maximum dissolved Si data in each column in Nozaki et al. (1999), Piepgra and Jacobsen (1988), Shimizu et al. (1994), and Zhang and Nozaki (1996) are used when plotting the observed data. Otherwise, the Si concentration data at the depth of 3000 m in Bruland and Lohan (2006) are used.

3.6. Scavenging of REEs by Carbonate Phases

The vertical profile of dissolved REEs in the Bering Sea has not been reported. Piepgras and Jacobsen (1988, 1992) reported the REE vertical profiles, along with silica data, at a station (TSP47: 39-1, 47°00'N, 161°08'E) in the North Pacific Ocean, which is the nearest station studied to the station in the Bering Sea (AB 53.5°N, 177°W) and whose silica concentration (maximum 178.2 µmol) approaches that in the Bering Sea (180-190µmol). Unfortunately, their data is not for dissolved REEs and includes particulate phases (Piepgras and Jacobsen, 1992), but the contribution of particulate phases should be about a few % of the dissolved REEs (Alibo and Nozaki, 1999), except for MREEs in deep water and Ce. The reported vertical profiles of La, Gd, and Lu are compared in Fig 6 with those reconstructed for opal dissolution using the next equation.

$$REE_{recon} = REE_{surface} + \left(\frac{REE}{Si}\right)_{opal} \times \Delta Si \qquad (3)$$

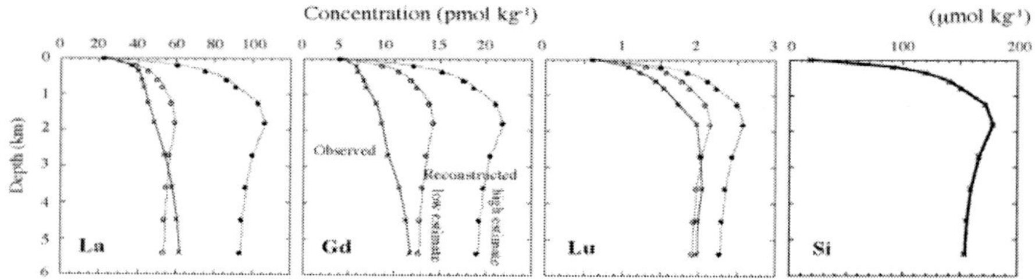

Figure 6. Comparison of the observed vertical profiles of La, Gd, and Lu (Piepgras and Jacobsen, 1992) with reconstructed profiles based on dissolve silica data (shown in the right, from Piepgras and Jacobsen, 1988) assuming that the REE composition of original diatom opal is given by the high estimate (filled circle) and the low estimate (open circle) (Table 1).

Opal with the compositions of the two estimates is considered. The diatom incorporation theory (Akagi, 2013) predicts that REE composition of diatom opal is governed by the dissolved silica concentration of deep water and that the difference in REE composition of diatom opal is very small between the Bering Sea and the North Pacific Ocean. Similarly to the discussion surrounding Fig. 5, it is readily noticeable that the dissolution of opal can essentially explain the increasing vertical profiles of REEs, but that dissolution provides more LREEs (La) and MREEs (Gd) than HREEs (Lu). We also noticed that the dissolution of opal could not supply enough La and Lu if the opal had the REE composition of the low estimate.

The difference between REE_{recon} and the observed REE concentration, REE_{obs}, may be explained by the adsorption of REEs onto some solid phases. Fig. 7a and 7b show the shale-normalized REE patterns of the difference ($REE_{recon}-REE_{obs}$) using the high estimate and the low estimate, respectively. When the high estimate is adopted as opal composition in eq. (3), the patterns have a similar distinctive triangular shape with a peak at Eu and a positive Ce anomaly, irrespective of depth (Fig. 7a). The resemblance of the triangular profiles with those of the acetic acid-soluble fractions of the settling particles has not escaped our attention and the patterns of the latter are also displayed in Fig. 7c. If the original diatom opal has an REE concentration lower than the high estimate, the difference patterns becomes irregular and variable, which is less similar to the pattern of acetic acid-soluble fraction, and if the low estimate is used instead, the shapes of the patterns become inconsistent (Fig. 7b). This consideration is one piece of supporting evidence for the high estimate as the appropriate composition of original diatom opal. Incidentally it is interesting to note that the difference of REEs becomes smaller at depths more than 1.5 km (Fig. 6), which may imply the dissolution of carbonate, especially aragonite, and the subsequent release of REEs under the depth.

In the previous paper the difference was not discussed in detail and the observed small difference in depth profiles between REEs was merely attributed to the differential rate of dissolution of each REE from diatom opal (Akagi et al., 2011). We now consider that the differential rate of dissolution (Akagi et al., 2011) does not play a major role in determining the depth profiles and that the acetic acid-soluble substance adsorbs or scavenges the REEs once released from diatom opal dissolution.

The partitioning pattern between the difference ($REE_{recon}-REE_{obs}$) and the total dissolved in seawater observed at each depth (Piepgras and Jacobsen, 1992) is illustrated in Fig. 8a and 8b. Again Fig. 8a and 8b are drawn employing the high estimate and low estimate as opal composition, respectively. When the high estimate is employed, a similar partitioning pattern is constantly seen irrespective of water depths from the shallow to the deep (Fig. 8a). The feature of the pattern includes a rather flat plateau in LREE region with a maximum at Sm, declining to HREE region and a positive Ce anomaly.

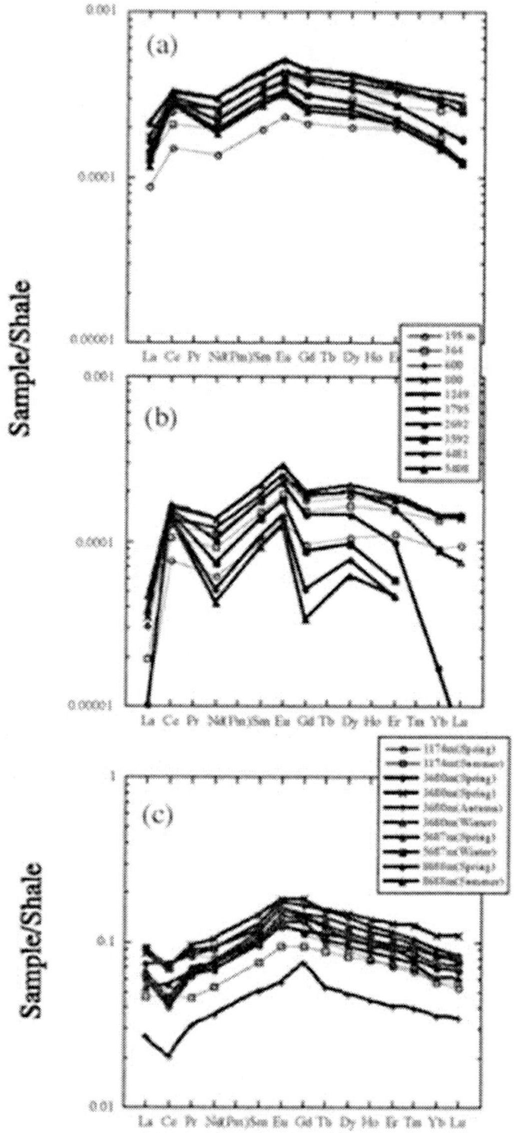

Figure 7. Shale normalized patterns of the discrepancies between the reconstructed profiles and the observed profiles (in g/g seawater basis) (see Fig. 3) at different depths adopting the high estimate (n=3) (a) and the low estimate (b) as opal composition. That of acetic acid-soluble fraction of suspended particles (in g/g particle basis, Lerche and Nozaki, 1998) (c). Refer to Lerche and Nozaki (1998) for the sample identification. Note that the discrepancy should be explained by scavenging of REEs once released through diatom dissolution.

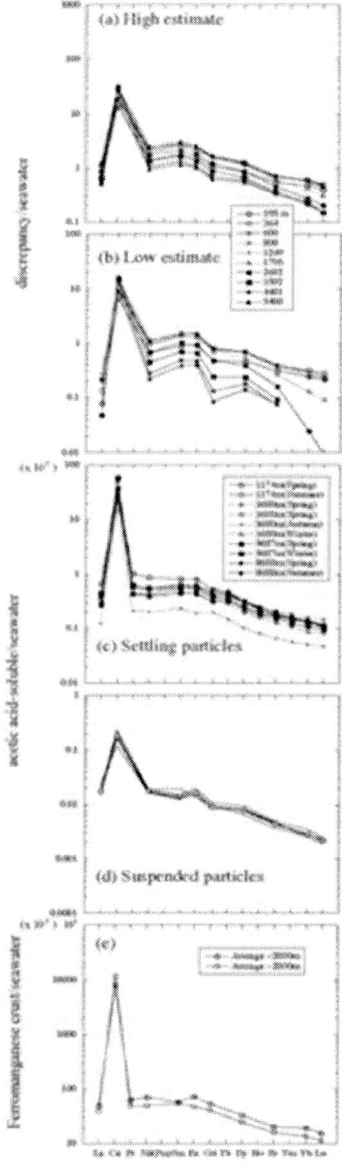

Figure 8. REE partitioning patterns against seawater for the discrepancies adopting the high estimate (n=3) (a) and the low estimate (b) as opal composition at different depths. (see Figs. 5 and 6), acetic acid-soluble fraction of settling particles (Lerche and Nozaki, 1998) (c) and acetic acid-soluble fraction of suspended particles (Sholkovitz et al., 1994) (d), and ferromanganese crusts (De Carlo et al., 1992) (e). Refer to individual literatures for the sample identification.

The pattern is also compared with the reported partitioning patterns between the acetic acid-soluble fraction of settling and suspended particles and that dissolved in seawater (Lerche and Nozaki, 1998; Sholkovitz et al., 1994) (Fig. 8c and d). The similarity among the three patterns indicates that the -acid soluble phase of particulate matter is responsible for the scavenging of dissolved REEs from seawater. Again if the low estimate is adopted as the opal composition, the consistency collapses (Fig. 8b) and partitioning patterns develop slumps at Gd and in the HREE region. This discussion provides a further piece of supporting evidence for the high estimate as the composition of original diatom opal.

The question is "What acetic acid-soluble substance scavenges REEs from seawater? Is it oxide, carbonate or anything else?" A similar partitioning pattern can be seen in manganese nodule/seawater (Fig. 8e). The partitioning reaction seems to be operative rather ubiquitously between solid and liquid in seawater columns. Carbonate has a rather monotonously decreasing pattern (Zhong and Mucci, 1995). But field carbonate data sometimes show partitioning patterns rather similar to our partitioning pattern (Scherer and Seitz, 1980). Iron or Fe-Mn oxyhydroxide can show LREE enriched partitioning patterns, if carbonate complexation with REEs is considered (Ohta and Kawabe, 2000a, b). Complex formation of REEs with carboxylate will also create a similar pattern (Byrne and Kim, 1990). Sorption experiments onto Mn and Fe oxides can also reproduce a similar partitioning pattern to a certain extent (Koeppenkastrop and De Carlo, 1992). The presence of Mn seems to be necessary to create a positive Ce anomaly (Ohta and Kawabe, 2001). Therefore, we consider it equally likely that oxide, carbonate, or organic matter can be responsible for the ubiquitous scavenging pattern.

3.7. New Calculation Method of Residence Time of REEs

The identification of diatom opal as the most important remover of REEs in surface water can provide a new method to estimate the residence time of REEs, relative to the turnover time of Si in seawater. In a water column dissolved silica is removed from the surface water by diatoms and is returned to the deep water by opal dissolution. The turnover time of dissolved silica, T_{Si}, is obtained by dividing the total amount of dissolved silica in the column by the intake of Si through diatom production per year. The turnover time of dissolved silica is not expected to be strongly affected by the dissolved silica concentration itself and we use it as a time standard in the calculation of the residence time of REEs. A turnover time of 405 years is obtained by dividing the total dissolved silica in seawater by the annual diatom intake rate of silica (Tréguer et al., 1995). It is not the same as the residence time of dissolved silica, which is as long as 15,000 years (Tréguer et al., 1995). Sedimentation of silica occurs in areas of very high production rate and in most of ocean columns the silica sedimentation rate is negligibly small compared with the turnover rate of silica.

In a system where an element is removed and supplied at an identical rate, the concentration of the element remains constant. If the removal rate suddenly increases with no change in the supply rate, the concentration of the element will be eventually decreased to adjust the removal rate to meet the supply rate. The residence time of the element is identical to the mass in the water column divided by the removal rate from the system. This consideration leads to the following equation to express the residence time of REE in a column, τ, as a function of turnover time of silica, T_{Si}. (see Appendix).

$$\tau = \frac{\left(\dfrac{\Delta REE_{obs}}{\Delta REE_{recon}} + \dfrac{REE_{surface}}{\Delta REE_{recon}} \right)}{\left(1 - \dfrac{\Delta REE_{obs}}{\Delta REE_{recon}} \right)} \times T_{Si} \tag{4}$$

Table 3 summarizes the results of calculation for each REE along with the reported values. Our residence time obtained using the high estimate as diatom opal composition (innate assumption) shows a contrasting variation from 190-270 years (Sm) to 1500-2100 years (Lu). These figures carry an uncertainty of approx. 20% owing to the uncertainty associated with the estimation of the turnover rate, T_{si}. The residence time of La (450-630 years) is longer than MREEs. The shorter residence time for MREEs is a result of more intensive scavenging by the acetic acid–soluble fraction of settling particles. Our residence time is largely consistent with reported values which ranges from 10^2 to 10^3 years (Table 2). The residence time of Nd has been intensively studied and our value (230-320 years) is one of the smallest values (Table 2). It should be noted that the residence time obtained in this manner may not represent those of global oceans, because the constituents (opal, carbonate and oxides) of settling particles, whose scavenging actions may be different from each other, vary from ocean to ocean.

Tasuku Akagi, Mariko Emoto, Rie Takada et al.

Table 3. Parameters and determination of mean residence time of rare earth elements in the ocean, compared with reported values (unit: year)

	La	Ce	Nd	Sm	Eu	Gd	Dy	Er	Yb	Lu
This study										
(Using high estimate composition)										
M_c/M_d*	0.31	0.051	0.22	0.18	0.20	0.26	0.31	0.42	0.35	0.36
M_d/M_d*	0.43	-0.01	0.27	0.24	0.27	0.36	0.41	0.56	0.68	0.74
$\tau_{actual\ system}$	450-630	14-20	230-320	190-270	220-310	340-480	430-600	770-1100	1100-1500	1500-2100
(Using low estimate composition)										
M_c/M_d*	0.72	0.10	0.41	0.31	0.32	0.46	0.48	0.59	0.47	0.46
M_d/M_d*	0.99	-0.02	0.50	0.41	0.42	0.63	0.63	0.77	0.92	0.95
$\tau_{actual\ system}$-	38000-54000	28-38	630-880	430-600	440-620	1000-1400	1000-1400	2100-2900	5900-8200	9300-13000
Earlier estimation										
(Budget observation)										
Elderfield and Greaves, 1982	240-690	110-400	450-620	220-600	460-470	300-720	340-810	420-980	410-970	
Piepgras and Wasserburg, 1983			200-300							
Tachikawa et al., 1999			200-1000							
Alibo and Nozaki, 1999	650-1630	50-130	380-950	390-970	320-820	440-1100	600-1510	2700-6800	Not estimated	Not estimated
(Model fitting)										
Tachikawa et al., 2003			487							
Shidall et al., 2008			500							
Arsouze et al., 2009			360							

The residence time has been estimated by dividing the total mass of each REE by the settling rate (Tachikawa et al., 1999) or input fluxes (Elderfield and Greaves, 1982; Nozaki et al., 1997; Alibo and Nozaki, 1999). Our estimation is based on the understanding of the role of each constituent and it is interesting to note that the values are fairly consistent with those obtained by an independent method, endorsing the validity of our understanding of the role particles play in water columns.

Incidentally, when the whole discussion is made using the low estimate of diatom opal (terrigenous assumption) in Table 2, the calculated residence time does not agree well with the reported values (Table 3). This is yet more evidence supporting for the suitability of the high estimate as the composition of diatom opal.

From the recent study of Nd isotope of diatom opal (Akagi et al., 2013), a new weathering process due to diatoms was discovered, which demands revision of the budgets of dissolved silica as well as rare earth elements. It is fortunate to us that our residence time is calculated based on the turnover rate of Si, not on the residence time that needs to be revised.

3.8. Diatom Incorporation Theory

Akagi (2013) developed the diatom incorporation theory. This was a sort of by-product, which had been introduced to explain the observed non-zero concentration of REEs in surface layers of the oceans. He pointed out that the stability constants of complexes of silicic acid with REEs were great enough for the complexes to dominate over conventional carbonate complexes. He introduced an assumption that diatoms incorporate REEs only in the form of silicic acid complexes and derived equations to quantify "leftover" (concentration ratio of surface and deep waters) and the concentration of REEs in surface water in terms of dissolved silica concentration of deep water. The equation successfully reproduced the observed REE concentrations of surface water in any ocean.

The diatom incorporation theory assumes that the sources of REEs are those dissolved in deep water, which was diffused/advected to the surface. The concentration of REEs in diatom opal can be expressed again as a function of dissolved silica concentration in deep water. The next equation shows the concentration of REEs in diatom opal:

$$\frac{-d[\text{REE}]}{-d[\text{Si}]} = a\,\frac{K_c\left(\dfrac{K_a}{[\text{H}^+]+K_a}\right)(1-\lambda)[\text{Si}_\text{T}]_{\text{DW}}}{1+K_c\left(\dfrac{K_a}{[\text{H}^+]+K_a}\right)(1-\lambda)[\text{Si}_\text{T}]_{\text{DW}}+\phi}. \tag{5}$$

In this equation, a is the empirical constant of concentration ratios of REEs and Si in deep water (see Akagi, 2013). K_a and K_c are acid dissociation constant of silicic acid and complex formation constant of a REE with silicic acid, respectively.

λ is the polymerization constant of silicic acid, which is negligibly small in surface water. (H^+) and $(Si_T)_{DW}$ are the concentration of H^+ in surface water and total dissolved silica in deep water, respectively. ϕ is the contribution of other complex formation, i.e.,

$$\phi = \beta_1\left[CO_3^{2-}\right] + \beta_2\left[CO_3^{2-}\right]^2 + K_{OH}\left[OH^-\right] + K_{SO_4}\left[SO_4^{2-}\right] + K_F\left[F^-\right] + K_{Cl}\left[Cl^-\right] + K_{PO_4}\left[PO_4^{3-}\right] \quad (6)$$

where each constant is the respective complex formation constant (Akagi, 2013).

The concentration of diatom opal calculated for the North Pacific Ocean using Eq. (5) corresponds to the *ideal* concentration of REEs in diatom opal, where diatoms absorb *REEs only from advected/diffused deep water*. In reality a portion of REEs is removed in a water column, which should somehow be supplied to keep the steady-state. This suggests that deep water is not the unique source of REEs.

In order to apply Eq. (5) to natural diatom opal, a correction for the implicit REE input is needed. If the input is assumed to be in balance with that removed (steady-state assumption), the factor for the correction should be the ratio of the amount of REEs from the diatom opal estimates to the amount corresponding to the observed increase in a depth profile.

Table 4 lists the produced values from Eq. (5), the correction factors assuming the higher and lower estimates, and compares the corrected values with both the estimates. Correction factors were calculated by integrating the reproduced profiles and reported profiles.

Again the REE and Si data reported for the North Pacific Ocean (TSP47: 39-1, 47°00'N, 161°08'E) by Piepgras and Jacobsen (1988, 1992) was used for the reproduction of the profiles and for drawing the reported profiles, respectively.

Both the estimate seems to be supported by the incorporation theory. However, it should be noted that correction factors for the low estimate is too small, which is not supported by the discussion in Section 3.5.

Table 4. Comparison of REE data between theoretical values (µg/g) of diatom opal predicted the diatom incorporation theory and the two estimates.

	La	Ce	Pr	Nd	Sm	Eu	Gd	Tb	Dy	Ho	Er	Tm	Yb	Lu
Theoretical values[a]														
Opal concn.	0.58	0.0997	0.0994	0.46	0.0967	0.0245	0.141	0.0251	0.163	0.0472	0.143	0.0241	0.173	0.0287
Correction factor (calculated using the integrated profile)[b]														
High estimate	2.33	-	-	3.71	4.20	3.71	2.76	-	2.24	-	1.79	-	1.48	1.35
Low estimate	1.02	-	-	2.00	2.42	2.37	1.59	-	1.59	-	1.29	-	1.09	1.06
Corrected Theoretical Values[c]														
High estimate	1.35	-	-	1.71	0.406	0.091	0.389	-	0.397	-	0.258	-	0.258	0.039
Low estimate	0.589	-	-	0.922	0.234	0.058	0.224	-	0.260	-	0.185	-	0.188	0.030
Average diatom opal[d]														
High estimate	1.08±15	2.4±4	0.30±5	1.13±17	0.26±2	0.061±1	0.26±3	0.041±8	0.26±6	0.056±12	0.18±4	0.027±4	0.19±4	0.032±5
Low estimate	0.47±6	1.2±1	0.157±7	0.61±1	0.15±2	0.039±3	0.16±5	0.027±3	0.17±2	0.038±7	0.13±3	0.020±2	0.14±2	0.025±3

a) Values are calculated using Eq. (5) at pH=7.8 and $(Si)_{DW}$=180 µmol/kg. b) Correction factors for the scavenged removal to keep the steady-state of the water column. c) Product of the theoretical value and the correction factor. d) The average data are the geometric mean of the three data (± one σ). In the high estimate, uncorrected values are adopted assuming that all the detected Al is innate (innate assumption) and in the low estimate the values are corrected assuming that the Al is from terrigenous contamination (terrigenous estimate).

Conclusion

The application of dissolution kinetics to the siliceous fractions of settling particles collected in an extremely diatom-productive area avoids the influence of alteration during the settlement, and confines the rare earth element composition of diatom opal to a value between the two estimates: the high estimate, which allows the presence of Al and the low estimate, which forbids its presence in diatom opal. We have proposed the high estimate for the typical diatom opal composition in the Bering Sea for the following reasons:

i   The presence of REEs in diatom opal after correction for terrigenous contamination invalidates the assumption of Al exclusion from opal.

ii   The dispersed presence of Al in any position of diatom frustules is observed by the SEM images.

iii   The difference between the dissolved REE composition of seawater reproduced by the dissolution of diatom opal having the composition of the high estimate and that observed are always positive in any depths, and consistent partitioning patterns are observed between the difference and REEs dissolved in seawater.

iv   A box model consideration using the high estimate gives residence times closer to those reported.

v   The incorporation theory can reproduce the composition.

The present paper shows that diatom opal should not be regarded as pure hydrated silicate, but as impure matter, which transports some metals to the deep water. Because of their minute size and sticky hydrous silica surface, it has long been taken for granted that diatoms are mixed with aluminosilicate particles. To separate opal from aluminosilicate, alkaline sodium carbonate treatment has conventionally been employed, but it should be noted that this conventional practice should automatically bias its true chemical image.

Appendix

A.1 Calculation of Residence Time Relative to the Turnover Rate of Dissolved Silica

Silica has a much longer residence time in ocean than REEs (e.g., τ_{Si}=15,000 years by Tréguer et al., 1995), and the sedimentation of opal only occurs in a limited area of extremely high productivity (Nelson et al., 1995). Dissolved silica is assumed to recycle without any loss by the time when REEs have been removed from the column. If REEs in opal behave identically to silica, the observed vertical profiles of REEs should be identical to that of the vertical profiles reconstructed from opal data. The difference is a result of selective removal of REEs. First let us consider a case where no element has been removed (system A; Fig. A1 a).

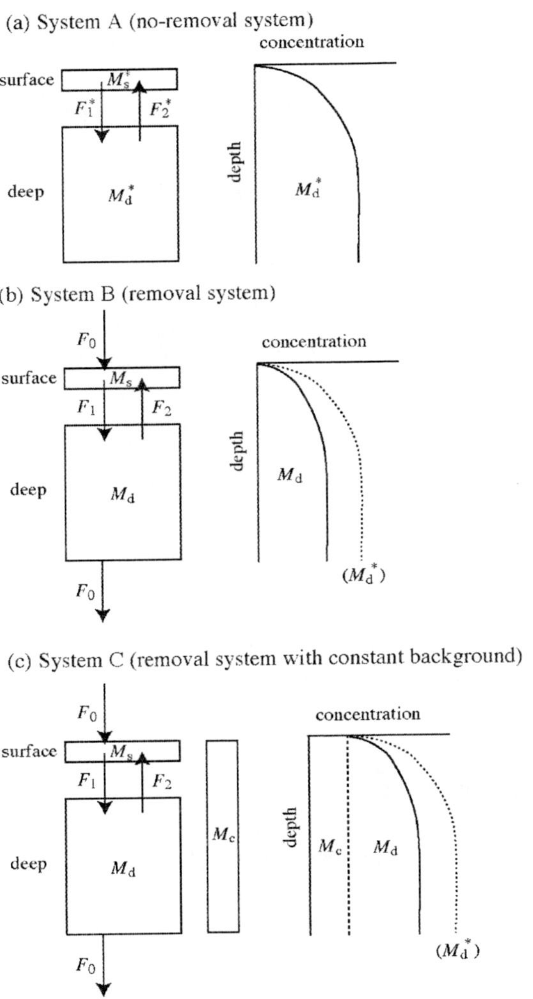

Figure A1. Schematic diagrams of box models for system A, no removal system (a), system B, removal system (b) and system C, removal system with constant background (c).

The vertical profile of the element is identical to that calculated from Si data as shown in the right of Fig. A1 a. In the box model this case is expressed in the left of Fig. A1 a. Identical fluxes from the surface box to the deep box and from the deep box to the surface box cancel each other. Because of no influx to or efflux from the whole system, τ_{system}, is infinity. (The asterisk stands for no removal-system.)

$$\tau^*_{system\ A} = \frac{M^*_{total}}{0} = \infty, \tag{A1}$$

where M^*_{total} is the sum of mass of element in the surface and deep boxes, M^*_s and M^*_d.

Residence time of element in the deep box, τ_d, is approximated to turnover rate of Si, T_{Si}, as the size of the surface box is much smaller than that of the deep box.

$$\tau^*_d = \frac{M^*_d}{F^*_1} = \frac{M^*_d}{F^*_2} \tag{A2}$$

$$\approx \frac{M^*_{total}}{F^*_1} = T_{Si} \tag{A3}$$

Let us consider the case where a part of element is removed from the deep box to out of the system (removal system; Fig. A1 b) with F1 being the same, i.e.,

$$F_1 = F^*_1. \tag{A4}$$

The sum of the two effluxes from the deep box is equal to the influx to the deep box at the steady state and another influx to the system, F_0, is implicitly necessary to keep the system at steady state.

$$F_1 = F_2 + F_0 \tag{A5}$$

The mass of element in the deep box decreases and the flux from the deep box to the surface box decreased accordingly, if the system acts in the same manner as the no-removal system with respect to water mass exchange and diffusion.

$$\frac{M_d}{F_2} = \frac{M_d^*}{F_2^*}$$

(A6)

The residence time of the removal system (system B) is

$$\tau_{system\ B} = \frac{M_{total}}{F_0} \approx \frac{M_d}{F_0}.$$

(A7)

This is further rewritten using eqs (A2), (A3), (A4), (A5) and (A6).

$$\tau_{system\ B} \approx T_{Si} \times \frac{M_d / M_d^*}{1 - M_d / M_d^*}$$

(A8)

In the case of actual surface water (system C), the concentration is not zero. This requests a following modification. By separating the vertical profile to two components (constant component and silica depending component) (Fig. A1 c),

$$\frac{M_c + M_d}{\tau_{system\ C}} = \frac{M_c}{\tau_c} + \frac{M_d}{\tau_{system\ B}},$$

(A9)

where τ_c means the residence time of the element in a new box for the constant component. In the constant box the residence time of the element is infinitive with a void of any in- and effluxes. Therefore,

$$\frac{M_c + M_d}{\tau_{system\ C}} \approx \frac{M_d}{\tau_{system\ B}}.$$

(A10)

Finally, using (A7) we obtain

$$\tau_{\text{system C}} \approx T_{\text{Si}} \times \frac{M_c/M_d^* + M_d/M_d^*}{1 - M_d/M_d^*}. \tag{A11}$$

Acknowledgment

This research was supported by MEXT/JSPS Grant Number 23310008.

References

Akagi, T., Fu F.-F., Hongo, Y. & Takahashi, K. (2011) Composition of Rare Earth Elements in Settling Particles Collected in the Highly Productive North Pacific Ocean and Bering Sea: Implications for Siliceous-Matter Dissolution Kinetics and Formation of Two REE-Enriched Phases. *Geochim. Cosmochim. Acta.* 75, 4857-4876.

Akagi, T. (2013) Rare earth element (REE)-silicic acid complexes in seawater to explain the incorporation of REEs in opal and the "leftover" REEs in surface water. *Geochim. Cosmochim. Acta.* 113, 174-192.

Akagi T., Yasuda S., Asahara Y., Tanimizu M, Emoto M. & Takahashi K. Diatoms spread a high εNd-signature in the Oceans.

Alibo D.S. and Nozaki Y. (1999) Rare earth elements in seawater: Particle association, shale-normalization, and Ce oxidation. *Geochim. Cosmochim. Acta.* 63, 363-372.

Arsouze T., Dutay J.-C., Lacan F. & Jeandel C. (2009) Reconstructing the Nd oceanic cycle using a coupled dynamical-biogeochemical model. Biogeosciences 6, 2829-2846.

Beck, L., Gehlen, M., Flank, A.-M., Van Bennekom, A.J., Van Beusekom, J.E.E. (2002) The relationship between Al and Si in biogenic silica as determined by PIXE and XAS. *Nuclear Instruments and Methods in Physics Research B* 189, 180–184.

Bruland, K. W. and Lohan M. C. (2006) Controls of trace metals in seawater. In *The Ocean and Marine Geochemistry* (ed. H. Elderfield) Vol 6 *Treatise on Geochemistry*, Elsevier-Pergamon, Oxford.

Byrne R. H. and Kim K.-H. (1990) Rare earth element scavenging in seawater. *Geochim. Cosmochim. Acta* 54, 2645-2656.

De Baar H.J.W., Bacon M.P. and Brewer P.G. (1985) Rare earth elements in the Pacific and Atlantic Oceans. *Geochim. Cosmochim. Acta.* 49, 1943-1959.

De Carlo E. H. and McMurtry G. M. (1992) Rare-earth element geochemistry of ferromanganese crusts from the Hawaiian Archipelago, central Pacific. *Chem. Geol.* 95, 235-250.

de la Rocha, C. L. (2006) The biological pump In *The Ocean and Marine Geochemistry* (ed. H. Elderfield) Vol 6 *Treatise on Geochemistry*, Elsevier-Pergamon, Oxford.

Ding, Z. L., Sun, J. M., Yang, S. L., & Liu, T. S. (2001) Geochemistry of the Pliocene red clay formation in the Chinese Loess Plateau and implications for its origin, source provenance and paleoclimate change. *Geochim. Cosmochim. Acta* 65, 901–913.

Dixit, S., Van Cappellen, P. & van Bennekom, A. J. (2001) Processes controlling solubility of biogenic silica and pore water build-up of silica acid in marine sediments. *Mar. Chem.* 73, 333-352.

Edmond J.M., Jacobs S.S., Gordon A.L., Mantyla A.W. and Weiss R.F. (1979) Water column anomalies in dissolved silica over opaline pelagic sediment and the origin of the deep silica maximum. *Jour. Geophys. Res.* 84, 7809-7826.

Elderfield H. and Greaves M.J. (1982) The rare earth elements in seawater. Nature 296, 214-219.

Ginoux, P., Chin, M., Tegen, I., Prospero, J., Holben, B., Dubovik, O., Lin, S.J. (2001) Sources and global distributions of dust aerosols simulated with the GOCART model. *J. Geophys. Res.* 106, 255–273.

Jahn, B., Gallet, S. & Han, J. (2001) Geochemistry of the Xining, Xifeng and Jixian sections, Loess Plateau of China: eolian dust provenance and paleosol evolution during the last 140 ka. *Chem. Geol.* 178, 71–94.

Kelemen, P. B., Hangoi, K. & Greene, A. R. (2003) One view of the geochemistry of subduction-related magmatic arcs, with an emphasis on primitive andesite and lower crust. In The Crust (ed. R.L. Rudnick) Vol. 3 Treatise on Geochemistry (Eds. H. D. Holland and K. K. Turekian), Elsevier-Pergamon, Oxford.

Koeppenkastrop, D. and De Carlo, E. H. (1992) Sorption of rare-earth elements from seawater onto synthetic mineral particles: An experimental approach. *Chem. Geol.* 95. 251-263.

Lerche D. and Nozaki Y. (1998) Rare earth elements of sinking particulate matter in the Japan Trench. *Earth Planet. Sci. Lett.* 159, 71-86.

McLennan S. M. (1989) Rare earth elements in sedimentary rocks: Influence of provenace and sedimentary processes. *Rev. Mineral.* 21, 169 –200.

Nelson D. M., Tréguer P., Brzezinski M. A., Leynaert A. and Quéguiner B. (1995) Production and dissolution of biogenic silica in the ocean: Revised global estimates, comparison with regional data and relationship to biogenic sedimentation. *Global Biogeochem. Cycle* 9, 359-372.

Nozaki Y., Alibo D.S., Amakawa H., Gamo T. and Hasumoto H. (1999) Dissolved rare earth elements and hydrography in the Sulu Sea. *Geochim. Cosmochim. Acta.* 63, 2171-2181.

Nozaki Y., Zhang J. and Amakawa H. (1997) The fractionation between Y and Ho in the marine environment. *Earth Planet. Sci. Lett.* 148, 329-340.

Ohta, A. and Kawabe, I. (2000a) Rare earth element partitioning between Fe oxyhydroxide precipitates and aqueous NaCl solutions doped with $NaHCO_3$: determinations of rare earth element complexation constants with carbonate ions. *Geochem. J.* 34, 349-454.

Ohta, A. and Kawabe, I. (2000b) Theoretical study of tetrad effects observed in REE distribution coefficients between marine Fe-Mn deposit and deep seawater, and in REE(III)-carbonate complexation constants. *Geochem. J.* 34, 455-475.

Ohta, A. and Kawabe, I. (2001) REE(III) adsorption onto Mn dioxide (δ-MnO_2) and Fe oxyhydroxide: Ce(III) oxidation by δ-MnO_2. *Geochim. Cosmochim. Acta.* 65, 695-703.

Oka A., Hasumi H., Obata H., Gamo T. & Yamanaka Y. (2009) Study on vertical profiles of rare earth elements by using an ocean general circulation model. Global Biogeochem. Cycles 23, GB4025, doi:10.1029/2008GB003353.

Piepgras D.J. and Jacobsen S.B. (1988) The isotopic composition of neodymium in the North Pacific. *Geochim. Cosmochim. Acta.* 52, 1371-1381.

Piepgras D.J. and Jacobsen S.B. (1992) The behavior of rare earth elements in seawater: precise determination of variations in the North Pacific water column. *Geochim. Cosmochim. Acta.* 56, 1851-1862.

Piepgras D. J. and Wasserburg G. J. (1983) Influence of the Mediterranean outflow on the isotopic composition of neodymium in waters of the North Atlantic. *Jour. Geophys. Res.* 88. 5997-6006.

Sambrotto R.N., Goering J.J. and McRoy C.P. (1984) Large yearly production of phytoplankton in the western Bering Strait. *Science* 225, 1147-1150.

Scherer, M. and Seitz, H. (1980) Rare-earth element distribution in Holocene and Pleistocene coral and their redistribution during diagenesis. *Chem. Geol.* 28, 279-289.

Shemesh, A., Mortlock, R.A., Smith, R.J., Froelich, P.N. (1988) Determination of GerSi in marine siliceous microfossils: sep- aration, cleaning and dissolution of diatoms and radiolaria. *Marine Chemistry* 25, 305–323.

Shimizu, H., Tachikawa, K., Masuda, A., Nozaki, Y. (1994). Cerium and neodymium isotope ratios and REE patterns in seawater from the north Pacific-Ocean. *Geochim. Cosmochim. Acta* 58, 323–333

Sholkovitz E.R., Landing W.M. and Lewis B.L. (1994) Ocean particle chemistry: The fractionation of rare earth elements between suspended particles and seawater. Geochim. Cosmochim. Acta. 58, 1567-1579.

Siddall, M., Khatiwala, S., van de Flierdt, T., Jones K., Goldstein, S.L., Hemming, S., & Anderson, R. F. (2008) Towards explaining the Nd paradox using reversible scavenging in an ocean general circulation model. *Earth Planet. Sci. Lett.* 274, 448-461.

Tachikawa K., Jeandel C. and Roy-Barman M. (1999) A new approach to the Nd residence time in the ocean: the role of atmospheric inputs. *Earth Planet. Sci. Lett.* 170, 433-446.

Tachikawa K., Athias V. and Jeandel C. (2003) Neodymium budget in the modern ocean and paleo-oceanographic implications. *J. Geophys. Res.* 108, NO. C8, 3254, doi:10.1029/1999JC000285.

Takahashi K., Fujitani N. and Yanada M. (2002) Long term monitoring of particle fluxes in the Bering Sea and the central subarctic Pacific Ocean, 1990-2000. *Progress in Oceanogr.* 55, 95-112.

Tréguer P., Nelson D. M., Van Bennekon A. J., DeMaster D. J., Leynaert A., Quéguiner B. (1995) The ilica balance in the world ocean: a reestimate. Science, 268, 375-379.

Tsunogai S., Kusakabe M., Izumi H., Koike I. and Hattori A. (1979) Hydrographic features of the deep water of the Bering Sea – the sea of silica. *Deep-Sea Res.* 26/6A, 641-659.

Zdanowicz, C., Hall, G., Vaive, J., Amelin, Y., Percival, J., Girard, I., Biscaye, P., & Bory, A. (2006) Asian dustfall in the St. Elias Mountains, Yukon, Canada. *Geochim. Cosmochim. Acta* 70, 3493–3507.

Zhang J. and Nozaki Y. (1996) Rare earth elements and yttrium in seawater: ICP-MS determinations in the East Caroline, Coral Sea and South Fiji basins of the western South Pacific Ocean. *Geochim. Cosmochim. Acta.* 60, 4631-4644.

Zhong, S. and Mucci, A. (1995) Partitioning of rare earth elements (REEs) between calcite and seawater solutions at 25°C and 1 atm, and high dissolved REE concentrations. *Geochim. Cosmochim. Acta.* 59, 443-453.

In: Diatoms
Editor: Flaubert C. Bour

ISBN: 978-1-62948-210-1
© 2013 Nova Science Publishers, Inc.

Chapter II

Diatom Flora of Fresh and Brackish Water Bodies of the Sakhalin Island (Far East, Russia)

T. V. Nikulina[*]
Institute of Biology and Soil Science,
Far East Branch of the Russian Academy of Sciences, Vladivostok, Russia

Abstract

The results of diatom flora investigation from water bodies of the Sakhalin Island are presented herein. Sakhalin, the biggest island of Russia, is washed by the Sea of Okhotsk and the Sea of Japan. It is separated by the Strait of Tartary from continental Asia (Russia) and by the La Pérouse Strait from the Hokkaido Island (Japan). Water bodies of Sakhalin (rivers, springs, fresh and brackish water lakes, hot spring) are represented by variety of types and sizes, and differ from each other by water temperature, pH, mineralization and trophic level. As a result, the diversity of diatom algae on Sakhalin is quite high. Similarities and differences of dominant diatom taxa and community composition between each type of water bodies are investigated and presented in this chapter. Over of 500 species, varieties and forms of diatoms from three

---

[*] Corresponding author: e-mail (nikulina@biosoil.ru, nikulinatv@mail.ru).

classes (Coscinodiscophyceae, Fragilariophyceae, and Bacillariophyceae) are recorded for the island. New records of diatoms for water bodies of the Sakhalin Island are reported. Ecological and geographical characteristic (relation to habitat, salinity, pH, saprobity, and geographical distribution) of the Sakhalin diatom flora are discussed.

Introduction

Investigation of diatom algae from fresh and brackish water bodies of the Sakhalin Island has been carried out by Russian scientists for more than 50 years. Data of these investigations have been published in numerous scientific journals. We find it necessity to summarize all available data on diatoms of the Sakhalin Island and then to make a taxonomic revision of the diatom flora.

To evaluate the current knowledge of diatoms of Sakhalin is the first step in a bigger project – to revise algal flora of the Russian Far East. Data on diatoms of both fresh and brackish water bodies can be used for comparison with the algal flora of the adjacent territories, as well as, to define significance of algal flora of this region. In addition, the hydro-biological research with the purpose to assess sanitary-biological status of water objects is based on species identification of algae as indicators of water quality.

Description of the Sakhalin Island and Its River Systems

Sakhalin, the largest island of Russia, is located near the east coast of Asia. Sakhalin is surrounded by the Sea of Okhotsk in the north and east, separated in the west from the mainland by the straits of Tartary and Nevelskoy, Amur Estuary and Sakhalin Bay, and from Hokkaido Island (Japan) by the La Pérouse Strait in the south.

The island is stretched longitudinally, being 948 km long, from the Cape of Elizabeth in the north to the Cape Crillon in the south. The maximal width of the island is 160 km, the minimal (at Poyasok Isthmus) is 26 km. The island area is around 78,000 km^2 (Resources..., 1973).

Sakhalin is characterizes by various relief with mountains and lowlands. The Western Sakhalin Mountains, a main island ridge, divides the river system of Sakhalin into two groups; the one of them belongs to the Sea of Okhotsk, the other – to the Sea of Japan. Coastline of Sakhalin is weakly indented; large gulfs of Aniva and Patience (Zaliv Terpenia) are located only in southern and middle parts of the island.

Climate of Sakhalin is severe. Winter last for 5–7 months with a lot of snow and low temperatures (sometimes down to –48 C°). Summer is short and rainy (2–3 months). The average air temperature of the coldest month (January) is –23 C° in the north of the island and –8 C° in the south. The average air temperature of the warmest month (July) doesn't exceed 15–18 C°.

Sakhalin river system includes more than 61178 streams. Most of them (98 %) are small creeks and rivers with length less than 10 km. The main rivers Tym (330 km length) and Poronai (350 km length) flow in central part of the island in longitudinal direction. All other streams flow in latitudinal direction and have shorter length. The average river density is 1.3 km/km^2. By type of current velocity streams in Sakhalin can be divided into mountain, lowland, and mixed ones. Streams from west and east slopes of the Western Sakhalin Mountains belong to the mountain type, with current velocity 2.5–3.0 m/s. Streams from the North-Sakhalin Valley belong to the lowland type, with current velocity 1.5–2.5 m/s. Streams derive most of their water from melting snow during period from April to June and from rain, and also get underground water in July–August and November–March (Resources..., 1973).

Freshwaters in Sakhalin are weakly mineralized because of high contribution of underground water in discharge. The highest level of water mineralization is observed in winter and equals 100–150 mg/l. During spring flood caused by melting snow and also by floods caused by typhoons the mineralization decreases to nearly 50–60 mg/l. Water mineralization is lower in streams of northern and southern parts of Sakhalin. Streams from the central part of Sakhalin belong to hydrocarbon class, streams from the western coast belong to the hydrocarbon-chloride class (Ca, Na group), while the rivers from northern and southern parts of Sakhalin belong to the hydrocarbon-chloride class (Mg, Na group). Because of small water mineralization, streams have low water hardness, which varies from 0.2 to 1.5 mg/l.

There are 16120 lakes in Sakhalin. Most of them are located in the northern and south-eastern parts of the island. The largest lakes are Nevskoye (area 178 km^2), Tunaicha (174 km^2), and Bolshoye Vavaiskoye (44.1 km^2). Most lakes with freshwater but along the coastline there are brackish and salt lakes, connected to seas by narrow channels or straits (Resources..., 1973).

Literature Review

Even if nowadays scientists draw close attention to diatoms of the Sakhalin Island, most of their hydro-biological research has been conducted in the southern part of the island where economically important large lakes with fresh and brackish water are located (e.g. Tunaicha, Bolshoye Vavaiskoye, Maloye Vavaiskoye, Bolshoye Chibisanskoye, and Maloye Chibisanskoye).

The first information on algae from Sakhalin was published by T.F. Koptyaeva (1964) for the Vavayi lakes, where data on diatom species, their quantity and biomass were presented. Investigation of diatom flora of the Vavayi lakes was conducted for 40 years. By now extensive data on algae of the Vavayi lakes have been published (Nikulina, 2005a, b; Motylkova, Konovalova, 2008; Labay et al., 2010; Genkal et al., 2011). The algal flora of the Tunaicha Lake, a lake-like lagoon, with weakly salted water (2.2–12 ‰), is rather well investigated also. In 1977, a joint survey was conducted on the lake and later on the results, including species structure, qualitative parameters for periphyton, microbial plankton, phytoplankton, and zooplankton, were published (Usova et al., 1980). Twenty years later personnel from the Sakhalin Scientific Research Institute of Fisheries and Oceanography continued exploring the Tunaicha Lake (Samatov et al., 2002; Motylkova, Konovalova, 2003, 2012a; Konovalova, Motylkova, 2006; Kalganova, Gertsog, 2012). Some papers were focused on seasonal changes of phytoplankton, quantity, and biomass of algae from the Izmenchivoye Lake, located on the coast of the Sea of Okhotsk (Motylkova, Konovalova, 2010; Kalganova, Gertsog, 2012). The diatom flora of small mountain streams from the southern part of Sakhalin, flowing into the Sea of Okhotsk, the Strait of Tartary, and the Gulf of Aniva were published recently (Nikulina, 2005a, b; Motylkova, Konovalova, 2012b; Mogilnikova et al., 2013).

The results on diatom species diversity from the east coast of Sakhalin (the Dagi River) was published by Medvedeva (2013) and from hot springs located in the watershed of the Dagi River by Nikulina (2009a), and Nikulina, Kociolek (2011). Special aspects of structural organization and distribution of algal communities are investigated in the largest rivers of the island, such as Tym River (Mikishin, 2008; Konovalova, Motylkova, 2011; Nikulina, 2009b, 2011a, b) and Poronai River (Konovalova, Motylkova, 2008; Nikulina, 2009b).

Diatom flora of water bodies on the west coast of Sakhalin still remains poorly known. Only some data on species structure and quantitative characteristics of phytoplankton communities were published for the Sladkoye

Lake (155 km^2), which is important for fishing industry (Knjazev, Kalganova, 2000a, b; Motylkova, Konovalova, 2011). Also diatom species diversity was described for some small lakes (0.1–2.5 km length) and creeks (length less than 10 km) (Knjazev, Kalganova, 2000a; Medvedeva, Miski, 2011). The algal flora of the north part of Sakhalin is unknown.

Materials and Methods

Algal growths were scrubbed from the stones with a scalpel and a brush with rough bristles. Each sample was preserved with formaldehyde. Plankton was sampled using a Ruttner's bathometer. Algal samples were collected from water bodies of Sakhalin Island in 2001–2012. The most important sampling locations are shown on the Figure 1.

Before microscopic observation, the diatoms were ignited in hydrogen peroxide (Swift 1967) and mounted in Naphrax. Light microscopes "Alphaphot 2" (Nikon, objectives 40x/0,65 and 100x/1,25 oil) and "Axioskop 40" (Zeiss, objectives 40x/0,65 and 100x/1,25 oil) were used for observation. Permanent slides are held at the Institute of Biology and Soil Science (Far Eastern Branch of the Russian Academy of Sciences, Vladivostok, Russia). For identifying algae, we used monographs, surveys and key-books by the following authors: Krammer, Lange-Bertalot (1986, 1988, 1991a, b); Hartley et al. (1996); Lange-Bertalot, Genkal (1999); Krammer (2000, 2002, 2003); Lange-Bertalot (2001), and others.

The list of diatoms represents a compilation of published data complemented with original data produced by the author. Taxonomic nomenclature and classification follow Algaebase (Guiry, Guiry, 2012). Synonyms used in previous publications for Sakhalin are included in brackets, as well as data on frequency of taxa occurrence, relationships to habitat, water salinity, pH, saprobity, and geographical distribution.

To estimate the frequency of taxa occurrence we used the six point scale (Korde, 1956): 1 – solitary (1–5 cells in the slide); 2 – rare (10–15 cells in the slide); 3 – not infrequent (25–30 cells in the slide); 4 – frequent (1 cell in each row of the cover glass at magnification with immersion); 5 – very frequent (several cells under the same conditions); 6 – in bulk (several cells in each visual field under the same conditions). The taxa with frequency of occurrence of 6 were referred to as dominant. In case of the absence of a "true" dominant in an algal community, a taxon with maximal frequency of taxa occurrence considered to be "the dominant".

T. V. Nikulina

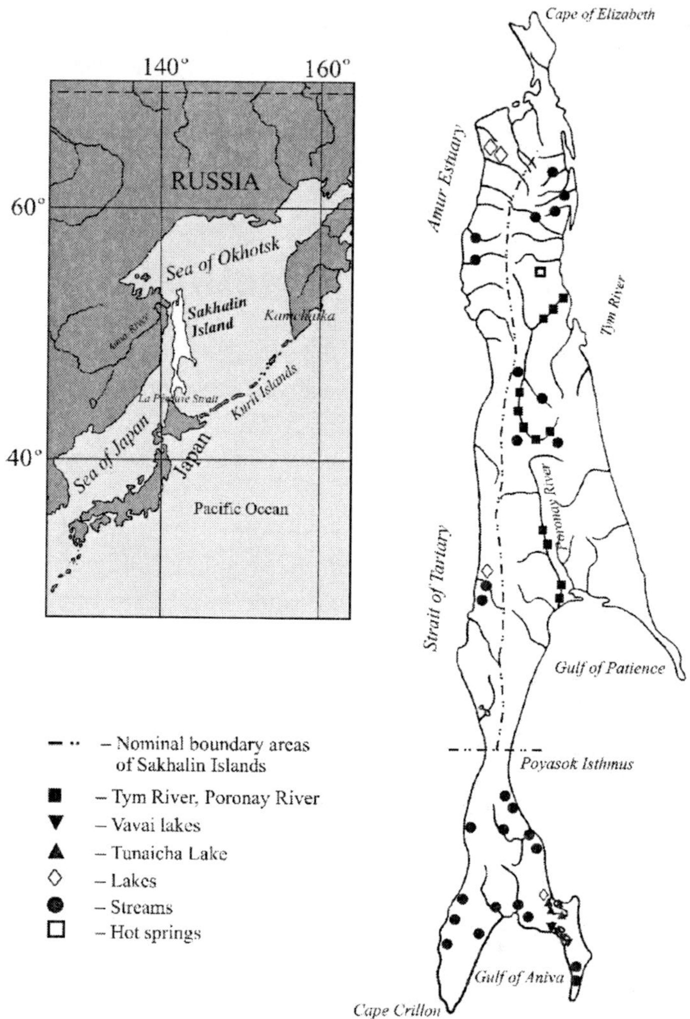

Figure 1. Map of sampling locations on the Sakhalin Island.

Taxonomic similarity measures of periphyton communities in water bodies on the east, west and south coasts of Sakhalin were estimated by means of cluster analysis, using the statistical program PAlaeontological Statistics, version 1.89 (Hammer et al., 2007). The Sörensen's coefficient was used as a similarity measure (Sörensen, 1948). The cluster-tracing algorithm used in the study was the Unweighted Pair Group Method with Arithmetic mean (UPGMA). Taxa for these analyses were identified to the lowest possible taxonomic level (i.e., down to form and variety).

For ecological and geographical analyses of algae we used publications of the following authors: Sladeček (1986); Van Dam et al. (1994); Bukhtiyarova (1999); Barinova et al. (2006).

Saprobity index was given on the basis of *Pantle-Buck method* (Pantle, Buck, 1955) in modification by Sladeček (1967).

Summary

The diatom flora of fresh and blackish water bodies of the Sakhalin Island includes 489 species (521 species, varieties, and forms) from three classes: Coscinodiscophyceae, Fragilariophyceae, and Bacillariophyceae (Tables 1, 2). Class Bacillariophyceae contains the largest number of taxa – 391 (75 % of total number of taxa). The family Cymbellaceae contains 69 taxa, followed by Bacillariaceae – 51, and Fragilariaceae – 50. In the taxonomic structure of the flora, the genus Pinnularia is most diverse and it includes 43 taxa, followed by Nitzschia – 36, Navicula – 32, *Eunotia* – 31.

The diatom flora of streams and lakes in the southern part of the island includes 367 species (391 species, varieties and forms) of algae from 99 genera. Water bodies of the southern Sakhalin with different parameters of water (temperature, pH, and mineralization, etc.), combined have high species richness of diatom algae and as a result complexes of dominant species are quite diverse.

Table 1. Taxonomic composition of diatom algae of the Sakhalin Island

Class	Order	Family	Genera	Species	Varieties and forms	Percent (%)
Coscinodiscophyceae	8	17	25	65	68	13.1
Fragilariophyceae	7	7	21	49	62	11.9
Bacillariophyceae	10	29	67	375	391	75.0
Total	25	53	113	489	521	100

Water in the large Vavayi lakes is weakly mineralized and belongs to the hydrocarbon class (Mg group). Temperature of water in summer warms up to 15.5–18.0 C° in Bolshoye Vavaiskoye Lake and Maloye Vavaiskoye Lake, and up to 18.6–21.5 °C in Chibisanskoe Lake, with average pH=6–7. Level of water mineralization fluctuates from 50 to 200 mg/l (Labai et al., 2010). *Aulacoseira granulata*, *A. italica*, and *Puncticulata comta* are abundant in plankton communities of the Vavayi lakes, whereas *Aulacoseira ambigua*, *Encyonema gracile*, *Rossithidium pusillum*, and *Cocconeis placentula* var. *euglypta* are dominat in periphyton communities.

Water in small freshwater lakes warms up to 18–23 C°; pH=6.5–7.0. *Staurosirella pinnata*, *Surirella linearis*, *S. robusta*, and *Aulacoseira valida* are most common in algal communities of those lakes. Water in lakes with blackish water (the Tunaicha Lake is the largest among them) warms up to 20–28.3°C in summer; pH fluctuates from 6.6 to 7.8; salinity level is 2.2 on the surface of water and 15.1 on the bottom (Samatov et al., 2002). Dominant complex of algae in the Tunaicha Lake is quite diverse with a number of benthic and benthic-plankton species: *Rhoicosphenia abbreviata*, *Diatoma vulgare*, *Gomphoneis olivaceum*, *Gomphonema parvulum*, *Melosira varians*, *Navicula cryptocephala*, *N. viridula*, *Cocconeis pediculus*, *C. placentula*, and *C. scutellum*.

Most abundant species in plankton and periphyton of the Izmenchivoye Lake, a lake-like lagoon, with salted water (25.8–32.8 ‰) are *Asterionella formosa*, *Cocconeis scutellum*, *Navicula gelida* var. *subimpressa*, *N. transitans*, *N. interglacialis*, *Odontella aurita*, *Thalassionema nitzschioides*, *Thalassiosira nordenskioeldii*, *Th. punctigera*, *Cyclotella meneghiniana*, and *Nitzschia frigida*.

Water in mountain streams of the southern part of the Sakhalin Island during summer warms up to 14.4–19.4 °C; pH=6.8–7.4; salinity is low in the headwaters (0.1–0.5 ‰) and rather high (10.8 ‰) in estuary (Mogilnikova et al., 2013). Algal flora of small streams is similar; however, complexes of dominant species are quite different. *Staurosirella pinnata*, *Planothidium lanceolatum*, *P. ellipticum*, *P. haynaldii*, *Achnanthidium minutissimum*, *Cocconeis placentula* var. *placentula*, *C. placentula* var. *euglypta*, *C. pediculus*, *Fragilaria vaucheriae*, *Fragilariforma virescens*, *Hannaea arcus*, *Meridion circulare*, *Diatoma mesodon*, *Ulnaria inaequalis*, *Rhoicosphenia abbreviata*, *Encyonema minutum*, *Gomphonema angustatum*, *Cymbopleura amphicephala*, *Hantzschia amphioxys*, *Nitzschia fonticola*, *N. palea*, and *N. paleacea* are most abundant on the surface of stones and other objects dip into water. *Ctenophora pulchella*, *Cocconeis scutellum*, and *Navicula phyllepta* are prevalent in estuary.

The diatom flora of the eastern part of Sakhalin includes 277 species (291 intraspecific species) of algae from 85 genera. Periphyton communities in streams flowing to the Sea of Okhotsk represents by dominant species, such as *Nitzschia palea, Tabellaria flocculosa, Diatoma mesodon, Didymosphenia geminata, Planothidium lanceolatum,* and *P. conspicuum.* Water temperature in these streams during summer varies in range 8.4–18°C; pH=6.2–7.9; electrical conductivity – 0.02; oxygen concentration – 9.89–11.49 mgO/l (Medvedeva, 2013).

Water of large rivers (Tym and Poronai) warms up during summer months up to 8.4–18 °C; pH=6.2–7.9; dissolved oxygen concentration varies from 4 to 12 mgO/l; concentration of minerals 30 to 130 mg/l (Resources ..., 1973). Composition of dominant diatoms in algal communities of these rivers of these rivers is different. *Encyonema silesiacum* is the main dominant species of the Tym River, from upstream to low section of the river. Two taxa, *Diatoma mesodon* and *Fragilaria capucina* var. *rumpens* are most common in the headwaters; *Hannaea arcus, Ulnaria inaequalis, Gomphonema parvulum, Cymbella affinis, C. tumida, Diatoma mesodon, Achnanthidium minutissimum, Melosira varians, Asterionella formosa, Planothidium lanceolatum, P. conspicuum,* and *Nitzschia brevissima* – in the low part of the river. *Melosira varians, Gomphoneis quadripunctatum, Hannaea arcus,* and *Ulnaria inaequalis* are most abundant in benthic algal communities of tributaries of the Tym River. Dominant species in benthic algal communities of the Poronai River are *Ulnaria inaequalis, Reimeria sinuata, Melosira varians, Cymbella tumida, Nitzschia brevissima, N. clausii,* and *N. obtusa.*

According to the published data, water of hydrothermal springs in the Dagi River watershed is characterized as being hydrocarbon-sodium chloride in composition, alkaline (pH=7–8), and water temperature varied from 20 to 55 °C (Karpunin et al., 1998, Zharkov, 2008). Diatom communities of periphyton and plankton in the hot springs have similar composition of species, but *Gomphonema angustatum, Anomoeoneis sphaerophora,* and *Tryblionella apiculata* were prevalent in periphyton samples, while *Achnanthes hungarica* was relatively abundant in plankton samples. The algal flora of Sakhalin hydrothermal springs is represented by common rheophilic diatoms indifferent to salinity (e.g. *Hannaea arcus, Gomphonema parvulum, Didymosphenia geminata, Synedra ulna, Encyonema silesiacum, Planothidium lanceolatum*), but number of maritime and brackish taxa is considerable.

Water temperature in lakes on the western coast of Sakhalin varies from 15.2 to 22°C; pH=6.4–7.2; concentration of oxygen – 5.5–5.6 mgO/l. The diatom flora of water bodies on the western coast of the island includes 190 species (203 species, varieties and forms) of algae from 59 genera. All year around in plankton samples from the Lake Sladkoye following taxa are most abundant: *Aulacoseira islandica*, A. granulata, and A. ambigua, whereas in benthic and plankton diatom communities of small lakes from the same area dominant taxa are represented by: *Aulacoseira islandica*, A. italica, Asterionella formosa, Diatoma mesodon, *Encyonema* silesiacum, Fragilaria capucina, Hannaea arcus f. arcus, H. arcus f. recta, and Navicula avenacea.

Small streams flowing from western slops of the Western Sakhalin Mountains are characterized by water temperature during summer from 9.3 to 15.6°C; pH=6.2–7.3. *Cocconeis* placentula, Cymbella cistula, Diatoma mesodon, Fragilaria capucina, *Fragilariforma* bicapitata, Hannaea arcus, Meridion circulare, M. circulare var. constrictum, Navicula avenacea, *Planothidium lanceolatum*, *P. haynaldii*, Achnanthidium minutissimum, and *Encyonema* silesiacum are most abundant in benthic communities of those streams.

Comparison of the diatom flora species composition of fresh and brackish water bodies belonging to the three conventional Sakhalin Island areas (south – the basin of the Sea of Japan, west – the Tartary Strait, and the Amur Estuary, east – the Sea of Okhotsk) by cluster analysis.

High degree of similarity of the flora of the southern, western and eastern coasts of the island was found (Figure 2). A single cluster is isolated in value of the index Dice = 0, 57 (Sörensen coefficient), it unites local diatom flora from water bodies of the basins Tartary Strait and the Amur Estuary. The low diatom species richness in streams and Sladkoe Lake on the western coast and high difference of species composition of lake ecosystems on the western and southern coasts explain the allocation of a single cluster. Recently Medvedeva and Miski (2011) published new records for the algal flora of Russia. They found diatom *Cymbella peraspera* var. *gigantea* and green *Stigeoclonium carolinianum* Islam in small lake and stream of the west coast of the Sakhalin Island. *C. peraspera* var. *gigantea* has been previously known from northern Germany and Romania, and *S. carolinianum* has been previously known from USA (North Carolina and Pennsylvania) and Israel.

Twelve diatoms are newly recorded in this chapter for the flora of the Sakhalin Island, these taxa are marked "*" in the Table 2.

Ecological analysis of the algal flora of Sakhalin revealed that data only for 85.4 % of total taxa are available. The majority of the identified taxa are benthic organisms (64.5 %) (Table 3). Relationship to water salinity is known for 352 taxa (67.5 %). The most abundant group is algae indifferent to water salinity (40.5 %). Data on relationship to water pH were found for 59.1 % of diatoms included in the survey. Among them alkaliphilous species (28.6 %) and indifferent to water pH species (17.3 %) are quite abundant (Table 3).

Type of geographical distribution is known for 64.1 % of total taxa recorded for the island. Over 40 % of species are cosmopolitan, 15.6 % – boreal, and 8.4 % arctic-alpine. Distribution for more than one third of the species could not be identified. The oligosaprobous (24.2 %) and betamesosaprobous (17.8 %) species are most common in the algal flora of the island (Table 3).

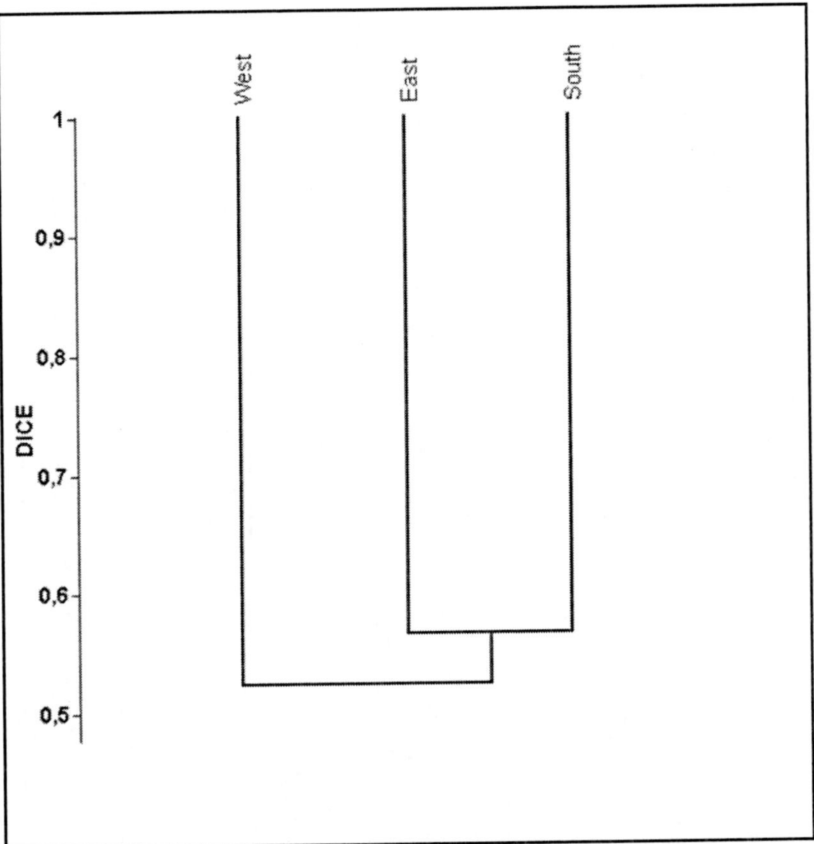

Figure 2. Dendrogram classification of water basins of the Sakhalin Island (on the vertical axis – the Sörensen coefficient, top – rivers basins: East – the Sea of Okhotsk, South – the Sea of Japan, West – the Amur Estuary, and the Strait of Tartary.

Table 2. Species composition of diatom flora of the Sakhalin Island

Taxa	Areas of Sakhalin Island			Ecological and geographical characteristic				
	East	South	West	B	H	pH	S	G
BACILLARIOPHYTA								
Class Coscinodiscophyceae								
Order Thalassiosirales								
Family Lauderiaceae								
Lauderia annulata Ceve (=*L. borealis* Gran)	-	+	-	-	-	-	-	-
Family Thalassiosiraceae								
Bacterosira fragilis Gran (=*B. fragilis* (Gran) Gran)	1	-	-	P	mh	-	-	a-a
Thalassiosira baltica (Grunow) Ostenfeld	-	+	-	-	hl	-	-	b
Th. bramaputrae (Ehr.) Håkansson et Locker (=*Th. lacustris* (Grunow) Hasle, *Coscinodiscus lacustris* Grunow)	+	1-3	+	-	-	-	β	-
Th. eccentrica (Ehrenberg) Cleve	1	-	-	P	mh	i	-	b
Th. gravida Cleve	1	-	-	P	mh	i	-	b
Th. hyalina Gran	-	+	-	P	-	-	-	-
Th. nativa Sheshukova-Poretzkaya	1	-	-	-	-	-	-	-
Th. nordenskioeldii Cleve	-	+	-	P	hl	-	-	-
Th. pacifica Gran et Angst	-	+	-	P	-	-	-	-
Th. proschkinae Makarova var. *spinulata* (Takano) Makarova	-	+	-	-	hl	-	-	-
Th. punctigera (Castracane) Hasle	-	+	-	P	-	-	-	-
Thalassiosira sp.	-	+	-	-	-	-	-	-
Family Skeletonemaceae								

Taxa	Areas of Sakhalin Island			Ecological and geographical characteristic				
	East	South	West	B	H	pH	S	G
Skeletonema costatum (Greville) Cleve	-	+	-	-	-	-	-	-
Family Stephanodiscaceae								
Cyclotella choctawhatcheeana Prasad (=*C. caspia* Grunow)	-	+	-	P	hl	-	-	-
C. meneghiniana Kützing (=*C. meneghiniana* Kützing var. *hankensis* Skvortzow, *C. kuetzingiana* Thwaites)	1	1-4	+	B-P	hl	alf	α-β	k
C. striata (Kützing) Grunow	1	-	-	-	hl	-	-	-
Cyclotella sp.	-	+	+	-	-	-	-	-
Cyclostephanos dubius (Fricke) Round	-	+	-	-	-	-	o-β	-
Discostella stelligera (Cleve et Grunow) Houk et Klee (=*Cyclotella stelligera* Cleve et Grunow)	+	1-3	+	-	-	-	o-β	-
Puncticulata comta (Ehrenberg) Håkansson (=*Cyclotella comta* (Ehrenberg) Kützing)	-	4-5	+	P	i	alf	β-o	k
P. radiosa (Lemmermann) Håkansson (=*C. radiosa* (Grunow) Lemmermann)	+	1-2	-	P	-	-	o-β	b
Stephanodiscus cf. *alpinus* Hustedt	+	+	+	-	-	-	-	-
S. delicatus Genkal	-	+	+	-	-	-	-	-
S. hantzchii Grunow	-	+	1-3	P	i	alf	α-β	k
S. makarovae Genkal	-	+	-	P	-	-	-	-
S. minutulus Kützing) Cleve et Möller	-	5	5	P	i	alf	o-β	k
Stephanodiscus sp.	-	+	-	-	-	-	-	-
Order Coscinodiscales								
Family Coscinodiscaceae								
Coscinodiscopsis commutata (Grunow) E.A. Sar & I. Sunesen	-	+	-	-	-	-	-	-

Table 2. (Continued)

Taxa	Areas of Sakhalin Island			Ecological and geographical characteristic				
	East	South	West	B	H	pH	S	G
Coscinodiscus asteromphalus Ehrenberg	+	-	-	P	-	-	-	-
C. concinnus W. Smith	-	+	-	P	-	-	-	-
C. jonesianus (Greville) Ostenfeld	-	+	-	P	-	-	-	-
C. marginatus Ehrenberg	+	-	-	P	-	-	-	-
C. oculus iridis Ehrenberg var. *oculus iridis*	+	-	-	P	-	-	-	-
C. oculus iridis var. *borealis* (Bailey) Cleve	-	1	-	P	-	-	-	-
Family Hemidiscaceae								
Actinocyclus octonarius Ehrenberg var. *octonarius* (=*A. ehrenbergii* Ralfs)	+	1	-	P	-	-	-	-
A. octonarius var. *ralfsii* (W. Smith) Hendey (=*A. ehrenbergii* var. ralfsii (W. Smith) Hustedt)	-	1	-	P	-	-	-	-
Family Heliopeltaceae								
Actinoptychus senarius (Ehrenberg) Ehrenberg (=*A. undulatus* (Bailey) Ralfs, *A. undulatus* var. *tamanica* Jousé)	+	1	-	-	-	-	-	-
Order Melosirales								
Family Melosiraceae								
Melosira arctica (Ehr.) Dickie	-	1	-	-	-	-	-	-
M. lineata (Dillwyn) Agardh	1	-	-	B-P	-	-	-	-
M. moniliformis (O. Müller) Agardh	-	+	-	B-P	hl	-	-	k
M. nummuloides (Dillwyn) Agardh (=*M. nummuloides* Agardh)	+	1-2	-	-	mh	-	-	-
M. varians Agardh	1-6	+	3-5	B-P	i	alb	β	k

Taxa	Areas of Sakhalin Island			Ecological and geographical characteristic				
	East	South	West	B	H	pH	S	G
Family Stephanopyxidaceae								
Stephanopyxis ferox (Greville) Ralfs	-	1	-	-	-	-	-	-
S. turris (Greville et Arnott) Ralfs	+	-	-	-	-	-	-	-
Family Hyalodiscaceae								
Hyalodiscus scoticus (Kützing) Grunow	+	-	-	-	-	-	-	-
Order Paraliales								
Family Paraliaceae								
Paralia sulcata (Ehrenberg) Cleve (=*Melosira sulcata* (Ehrenberg) Kützing)	1-3	1	-	P	mh	-	-	b
Order Aulacoseirales								
Family Aulacoseiraceae								
Aulacoseira alpigena (Grunow) Krammer	-	1-3	-	P	i	i	o	k
A. ambigua (Grunow) Simonsen	1	6	+	P	i	alf	α-β	k
A. distans (Ehrenberg) Simonsen	1	1-3	-	P	i	alf	α-β	k
A. granulata (Ehrenberg) Simonsen (=*Melosira granulata* (Ehrenberg) Ralfs)	1	1-6	+	P	i	alf	β	k
A. islandica f. *islandica* (O. Müller) Simonsen (=*Melosira islandica* O. Müller, *M. islandica* subsp. *helvetica* O. Müller)	-	+	+	P	i	acf	o-χ	b
A. islandica f. *curvata* (O. Müller) Simonsen (=*Melosira islandica* f. *curvata* O. Müller)	-	+	-	P	-	-	-	-
A. italica (Ehrenberg) Simonsen (=*Melosira italica* (Ehrenberg) Kützing, *A. italica* var. *tenuissima* (Grunow) Ehrenberg, *A. italica* var. *tenuissima* (Grunow) Ehrenberg)	1	1-5	1-6	P	i	alf	o-β	k
A. subarctica (O. Müller) Haworth	1-2	+	+	P	i	alb	α-β	k

Table 2. (Continued)

Taxa	Areas of Sakhalin Island			Ecological and geographical characteristic				
	East	South	West	B	H	pH	S	G
A. valida (Grunow) Krammer (=*Melosira italica* (Ehrenberg) Kützing var. *valida* (Grunow) Hustedt, *Aulacoseira italica* var. *valida* (Grunow) Simonsen)	-	3	-	P	i	alb	-	a-a
Aulacoseira sp.	2-3	-	1	-	-	-	-	-
Family Orthoseiraceae								
Orthoseira roeseana (Rabenhorst) O'Meara (=*Melosira roeseana* Rabenhorst, *Aulacosira epidendron* (Ehrenberg) Crawford)	+	-	-	B	i	alf	o	b
Order Triceratiales								
Family Triceratiaceae								
Cerataulus turgidus (Ehrenberg) Ehrenberg	1	-	-	-	-	-	-	-
Odontella aurita (Lyngbye) Agardth	1	1-2	-	B-P	mh	alf	-	-
Family Plagiogrammaceae								
Dimeregramma minor (Gregory) Ralfs	1	-	-	-	-	-	-	-
Order Rhizosoleniales								
Family Rhizosoleniaceae								
Dactyliosolen fragilissimus (Bergon) Hasle	-	+	-	-	-	-	-	-
Rhizosolenia fragilissima Bergon	-	+	-	P	-	-	-	-
Rh. setigera Brightwell	-	+	-	P	-	-	-	-
Order Chaetocerotales								
Family Chaetocerotaceae								
Chaetoceros diadema (Ehrenberg) Gran (=*Ch. subsecundus* (Grunow) Hustedt)	-	+	-	B-P	-	-	-	-

Taxa	Areas of Sakhalin Island			Ecological and geographical characteristic				
	East	South	West	B	H	pH	S	G
Ch. muelleri Lemmermann	-	+	-	B-P	hl	-	-	k
Ch. subtilis Cleve	-	+	-	B-P	-	-	-	-
Chaetoceros sp.	1	+	-	-	-	-	-	-
Class Fragilariophyceae								
Order Fragilariales								
Family Fragilariaceae								
Asterionella formosa Hassall (=*A. gracillima* (Hantzsch) Heiberg, *Asterionellopsis gracillima* (Hantzsch) Heiberg))	1	1-5	1-6	P	i	alf	o-β	k
Ctenophora pulchella (Ralfs ex Kützing) Williams et Round (=*Synedra pulchella* (Ralfs) Kützing)	1	2-6	-	B-E	mh	alf	β-α	b
Diatoma anceps (Ehrenberg) Kirchner	1-2	1	1-2	B	hb	alf	o-χ	a-a
D. ehrenbergii Kützing	1-2	-	-	B	-	-	o-β	k
D. hiemale (Roth) Heiberg	1-3	1-3	+	B	hb	i	χ	a-a
D. mesodon (Ehrenberg) Kützing (=*D. hiemale* var. *mesodon* (Ehrenberg) Grunow)	1-6	1-5	2-6	B	hb	alf	χ	a-a
D. moniliforme Kützing	1-2	1	-	B-P	hl	-	β-α	k
D. tenue Agardh (=*D. elongatum* (Lyngbye) Agardh	1-2	1-3	2-4	B-P	hl	i	β-α	k
D. vulgare Bory	1-2	1	-	B-P	i	alb	β	b
Fragilaria capucina Desmazières var. *capucina*	1-3	1	2-6	B-P	i	alf	o-β	k
F. capucina var. *amphicephala* (Kützing) Lange-Bertalot ex Bukhtiyarova (=*Synedra amphicephala* Kützing)	-	+	-	B	i	i	χ	k
F. capucina var. *gracilis* (Oestrup) Hustedt	1	-	5	-	-	-	β-α	-

Table 2. (Continued)

Taxa	Areas of Sakhalin Island			Ecological and geographical characteristic				
	East	South	West	B	H	pH	S	G
F. capucina var. *mesolepta* (Rabenhorst) Rabenhorst	2	-	4-5	B-P	i	alf	o-β	k
F. capucina var. *rumpens* (Kützing) Lange-Bertalot ex Bukhtiyarova	1-5	1	2	B	i	acf	o	k
F. crotonensis Kitton	-	1-2	+	P	i	alf	o-β	b
F. distans (Grunow) Bukhtiyarova (=*Synedra rumpens* Kützing var. *fragilarioides* Grunow)	-	1-3	5	B-P	i	i	o	k
F. exigua Grunow (=*F. construens* var. *exigua* (W. Smith) Schulz	1	1-3	-	-	-	-	-	β-α	-
F. radians (Kützing) D.M.Williams & Round*	-	-	1	B-P	-	-	-	-
F. tenera (W. Smith) Lange-Bertalot (=*Synedra tenera* W. Smith)	1-2	-	2	B-P	-	-	-	-
F. vaucheriae (Kützing) Petersen (*F. intermedia* Grunow)	1-4	1-3	2-6	E	i	alf	o-β	k
Fragilaria sp.	-	-	1	-	-	-	-	-
Fragilariforma bicapitata (A. Mayer) Williams et Round	1	+	4-6	B	hb	acf	o-β	b
F. constricta (Ehrenberg) Williams et Round f. *constricta*	1	1	+	B	i	acf	-	a-a
F. constricta f. *stricta* (A. Cleve) Hustedt	-	1	-	B	i	acf	-	b
F. virescens (Ralfs) Williams et Round	-	6	-	B	hb	i	χ	a-a
Hannaea arcus (Ehrenberg) Patrick var. *arcus* f. *arcus*	1-6	1-6	6	B	i	alf	χ	a-a
H. arcus var. *arcus* f. *recta* (Cleve) Foget	1-4	1-3	2-6	B	i	alf	χ	a-a
H. arcus var. *amphioxys* (Rabenhorst) Patrick	1-4	1-2	-	-	-	-	o-χ	-
H. arcus var. *linearis* (Holmboe) Ross f. *linearis*	1-4	1	2	P	i	alf	-	a-a

Taxa	Areas of Sakhalin Island			Ecological and geographical characteristic				
	East	South	West	B	H	pH	S	G
Martyana martyi (Héribaud) Round (=*Opephora martyi* Héribaud)	-	2-3	-	B	i	alf	o-α	k
Meridion circulare (Greville) Agardh var. *circulare*	1-5	1-6	3-6	B	hb	alf	χ-o	k
M. circulare var. *constrictum* (Ralfs) Van Heurck (=*M. circulare* var. *constricta* (Ralfs) Van Heurck)	1-2	1-2	4-6	B	hb	alf	χ-o	k
Pseudostaurosira brevistriata (Grunow) Williams et Round var. *brevistriata* (=*Fragilaria brevistriata* Grunow)	1	+	-	B-P	i	alf	χ-o	k
Staurosira construens Ehrenberg f. *construens* (=*Fragilaria construens* (Ehrenberg) Grunow)	1-2	2	2	B-P	i	alf	o	k
S. construens f. *venter* (Ehrenberg) Bukhtiyarova	1-4	1-3	4	B-P	i	alf	β-α	k
S. construens var. *binodis* (Ehrenberg) Hamilton	-	1	1	B-P	i	alf	-	k
S. construens var. *triundulata* (Reichelt) Bukhtiyarova (=*Fragilaria construens* var. *triundulata* Reichelt)	-	-	+	B-P	i	alf	-	k
Staurosirella berolinensis (Lemmermann) Bukhtiyarova (=*Synedra berolinensis* Lemmermann)	-	3	+	-	-	-	-	-
S. leptostauron (Ehrenberg) Williams et Round	1	-	1	B	-	-	o-β	-
S. pinnata (Ehrenberg) Williams et Round (=*Fragilaria pinnata* Ehrenberg)	1-2	1-3	5	B-P	hl	alf	β-α	k
S. pinnata var. *trigona* (Brun et Héribaud) Siver et Hamilton (=*Fragilaria pinnata* var. *trigona* (Brun et Héribaud) Hustedt)	-	-	+	-	-	-	-	-
Synedrella parasitica (W. Smith) Round et Maidana (=*Fragilaria parasitica* (W. Smith) Grunow)	-	1-2	-	B	i	alf	χ	k
S. subconstricta (Grunow) Round et Maidana (=*Synedra parasitica* (W. Smith) Hustedt var. *subconstricta* (Grunow) Hustedt, *S. parasitica* var. *subconstricta* Grunow)	-	-	+	E	i	alf	o-β	k

Table 2. (Continued)

Taxa	Areas of Sakhalin Island			Ecological and geographical characteristic				
	East	South	West	B	H	pH	S	G
Ulnaria acus (Kützing) Aboal (=*Fragilaria ulna* var. *acus* (Kützing) Lange-Bertalot, *Synedra acus* Kützing)	-	1	1-4	P	i	alb	o-α	k
U. danica (Kützing) Compère et Bukhtiyarova (=*U. ulna* var. *danica* (Kützing) Grunow)	1-2	1	-	B-P	i	alf	β	k
U. inaequalis (H.Kobayasi) M.Idei (=*Synedra goulardii* (Brébisson) Grunow)	1-6	1-6	+	B	-	-	-	a-a
U. oxyrhynchus (Kützing) Aboal	1	-	-	B	i	alf	β-α	k
U. ulna (Nitzsch) Compère var. *ulna*	1-3	1-2	2-5	B-P	i	alf	o-α	k
Tabularia fasciculata (Agardh) Williams et Round	1	-	-	B	hl	alf	χ-o	k
T. tabulata (Agardh) Snoeijs (=*Synedra tabulata* (Agardh) Kützing)	1	2-5	+	B	mh	i	α-β	k
Order Tabellariales								
Family Tabellariaceae								
Tabellaria fenestrata (Lyngbye) Kützing	1-4	1-5	1-5	B-P	hb	acf	β	b
T. flocculosa (Roth) Kützing (=*T. fenestrata* var. *intermedia* Grunow)	1-6	1-5	1-5	B-P	hb	acf	o-χ	a-a
Tetracyclus ellipticus (Ehrenberg) Grunow	-	+	-	-	-	-	-	-
T. rupestris (Braun) Grunow	-	1	-	B	i	-	χ-β	a-a
Order Licmophorales								
Family Licmophoraceae								
Licmophora abbreviata Agardh	-	+	-	-	-	-	-	-
Order Rhabdonematales								
Family Rhabdonemataceae								

Taxa	Areas of Sakhalin Island			Ecological and geographical characteristic				
	East	South	West	B	H	pH	S	G
Rhabdonema arcuatum (Lyngbye) Kützing	-	+	-	-	-	-	-	-
Rh. minutum Kützing	1	-	-	-	-	-	-	-
Order Rhaphoneidales								
Family Rhaphoneidaceae								
Delphineis surirella (Ehrenberg) Andrews	1	-	-	-	-	-	-	-
Order Thalassionemales Round								
Family Thalassionemataceae Round								
Thalassionema frauenfeldii (Grunow) Hallegraeff (=*Th. frauenfeldii* Grunow)	-	+	-	-	-	-	-	-
Th. nitzschioides (Grunow) Mereschkowsky (=*Thalassionema nitzschioides* (Grunow) Van HeurckTh. *nitzschioides* Grunow)	1	+	-	B-P	-	-	-	-
Order Striatellales								
Family Striatellaceae								
Grammatophora angulosa Ehrenberg var. *islandica* (Ehrenberg) Grunow	-	3	-	B	-	-	-	-
G. arcuata Ehrenberg	-	+	-	B	mh	-	-	b
Class Bacillariophyceae								
Order Eunotiales								
Family Eunotiaceae								
Eunotia arcus Ehrenberg	-	1	-	B	hb	acf	o	k
E. bidens Ehrenberg	1	-	-	B	-	-	-	-
E.bilunaris (Ehrenberg) Mills (=*E. lunaris* (Ehrenberg) Grunow)	1-2	1	2	B	i	acf	β	k
E. crista-galli Cleve*	-	-	1	B	i	acf	-	a-a
E. diadema Ehrenberg (=*E. serra* Ehrenberg var. *diadema* (Ehrenberg) Patrick)	-	1	-	B	-	acf	-	-

Table 2. (Continued)

Taxa	Areas of Sakhalin Island			Ecological and geographical characteristic				
	East	South	West	B	H	pH	S	G
E. diodon Ehrenberg	1	1	+	B	hb	acf	o-χ	a-a
E. exigua (Brébisson ex Kützing) Rabenhorst	1	+	-	B	i	acf	χ	k
E. flexuosa (Brébisson) Kützing	1	-	-	B	-	acf	-	-
Eunotia formica Ehrenberg	1	+	2	B	hl	i	o	k
E. glacialis Meister (=*E. gracilis* (Ehrenberg) Rabenhorst, *E. valida* Hustedt)	1	+	+	B	-	acf	ρ	k
E. implicata Nörpel, Lange-Bertalot & Alles	1-2	-	1	B	-	acf	-	-
E. intermedia (Krasske) Nörpel & Lange-Bertalot	-	2	-	B	-	acf	-	-
E. minor (Kützing) Grunow	1	-	-	B	-	-	χ	-
E. monodon Ehrenberg (=*E. monodon* var. *major* (W. Smith) Hustedt)	-	+	-	B	hb	acf	β-o	k
E. mucophila (Lange-Bertalot et Nörpel-Schempp) Lange-Bertalot (=*E. bilunaris* var. *mucophila* Lange-Bertalot et Nörpel-Schempp)	1	1	-	B	-	acf	-	-
E. muscicola var. *tridentula* Nörpel-Schempp et Lange-Bertalot (=*E. polydentula* Brun)	-	+	-	B	-	acf	o-β	-
E. parallela Ehrenberg var. *parallela*	-	+	-	B	i	acf	β-o	b
E. parallela var. *angusta* Grunow	1	-	-	B	-	-	-	-
E. pectinalis (Dillwyn? Kützing) Rabenhorst	1	1	-	B	hb	acf	χ	k
E. praerupta Ehrenberg	1-4	1	1	B	hb	acf	χ	k
E. polyglyphis Grunow	-	+	-	B	hb	acf	o-χ	a-a
E. revoluta Cleve	-	+	-	B	hb	-	o	a-a
E. septentrionalis Oestrup	1	-	-	B	hb	acf	o	a-a

Taxa	Areas of Sakhalin Island			Ecological and geographical characteristic				
	East	South	West	B	H	pH	S	G
E. serra Ehrenberg var. *serra* (=E. *robusta* Ralfs)	1	+	-	B	hb	acf	o-β	a-a
E. serra var. *tetraodon* (Ehrenberg) Nörpel (=*E. robusta* var. *tetraodon* (Ehrenberg) Ralfs)	-	+	-	B	hb	acf	β-o	a-a
E. soleirolii (Kützing) Rabenhorst	1	-	2	B	-	acf	-	-
E. subarcuatoides Alles, Nörpel & Lange-Bertalot	1	-	-	B	-	-	-	-
E. sudetica O. Müller	-	+	-	B-P	i	acf	o-β	b
E. tenella (Grunow) Hustedt	-	+	-	B	hb	acf	χ-o	k
E. veneris (Kützing) De Toni (=E. *incisa* W. Smith ex Gregory)	1-2	-	2	B	hb	acf	β-o	k
Eunotia sp.	-	+	-	-	-	-	-	-
Order Lyrellales								
Family Lyrellaceae								
Lyrella atlantica (Schmidt) Mann	-	1	-	-	-	-	-	-
Petroneis marina (Ralfs) Mann (=*Navicula punctulata* W. Smith var. *pagophila* Grunow)	1-2	1	-	B	eu	-	-	b
Order Mastogloiales								
Family Mastogloiaceae								
Mastogloia elliptica (Agardh) Cleve	1	-	-	B	mh	alf	-	k
M. exigua Lewis	-	1-2	-	B	-	-	-	-
M. smithii Thwaites	1	1	-	B	mh	alf	β	k
Order Cymbellales								
Family Rhoicospheniaceae								

Table 2. (Continued)

Taxa	Areas of Sakhalin Island			Ecological and geographical characteristic				
	East	South	West	B	H	pH	S	G
Rhoicosphenia abbreviata (Agardh) Lange-Bertalot (=Rh. *curvata* (Kützing) Grunow)	1	1	2-3	B	hl	alf	β	k
Family Anomoeoneidaceae								
Anomoeoneis sphaerophora (Ehrenberg) Pfitzer	1-5	-	-	B-P	hl	alb	χ-β	k
Staurophora amphioxys (Gregory) Mann (=*Stauroneis gregori* Ralfs)	-	1	-	B	-	-	-	-
S. wislouchii (Poretzsky & Anisimova) D.G.Mann (=*Stauroneis wislouchii* Poretzky & Anisimova)	-	+	-	B	mh	-	-	k
Family Cymbellaceae								
Brebissonia lanceolata (Agardh) Mahoney et Reimer (=*B. boeckii* (Ehrenberg) Grunow)	1-2	1	1	B	mh	-	-	b
Cymbella affinis Kützing	1-6	1	-	B	i	alf	o-β	b
C. amplificata Krammer	1	1	1	B	-	-	-	-
C. aspera (Ehrenberg) H.Peragallo	1-2	-	1	B	i	alf	β-o	k
C. cistula (Ehrenberg) Kirchner	1-3	1-3	2-6	B	i	alf	β	b
C. cymbiformis Agardh	-	1	-	B	i	alf	-	k
C. helvetica Kützing	-	-	1	B	i	alf	o-α	k
C. hustedtii Krasske	-	1-2	-	B	i	alf	o	k
C. lanceolata (Ehrenberg) Kirchner	-	+	-	B	i	alf	β	-
C. neocistula Krammer	1	-	-	B	i	alf	-	-
C. parva (W.Smith) Kirchner	1	1	+	B	i	i	-	b

Taxa	Areas of Sakhalin Island			Ecological and geographical characteristic				
	East	South	West	B	H	pH	S	G
C. perparva Krammer	1	-	-	B	-	-	-	-
C. stuxbergii (Cleve) Cleve (=*C. stuxbergii* Cleve)	-	1	-	B	i	-	-	a-a
C. tumida (Brébisson) Van Heurck	1-5	1	+	B	i	alf	o	b
C. turgidula Grunow	1	1	-	B	-	i	-	k
Cymbella sp.	-	+	-	-	-	-	-	-
Cymbopleura amphicephala (Nägeli) Krammer (=*Cymbella amphicephala* Nägeli)	-	+	-	B	i	i	o-β	b
C. angustata (W. Smith) Krammer	1	-	-	B	i	i	o	b
C. cuspidata (Kützing) Krammer (*Cymbella cuspidata* Kützing)	1	1	-	B	i	i	o-α	k
C. inaequalis (Ehrenberg) Krammer	-	-	3	B	-	-	-	-
C. naviculiformis (Auerswald) Krammer (=*Cymbella naviculiformis* Auerswald)	1	1	1-2	B	i	i	o	k
Encyonema caespitosum Kützing (*Cymbella caespitosa* (Kützing) Brun)	-	1-4	-	B	i	-	β-α	k
E. elginense (Krammer) Mann (=*Cymbella turgida* (Gregory) Cleve)	1	1	2-3	B	hb	acf	-	-
E. gracile Ehrenberg (=*Cymbella gracilis* (Ehrenberg) Kützing)	1	1-6	4	B	hb	i	β	a-a
E. mesianum (Cholnoky) D.G. Mann (=*C. mesiana* Cholnoky)	-	1	-	B	-	-	o	-
E. minutum (Hilse ex Rabenhorst) Mann	1-4	2-6	2-3	B	i	i	o	k
E. muelleri (Hustedt) Mann (=*Cymbella muelleri* Hustedt)	-	1	-	B	-	alb	-	-
E. prostratum (Berkley) Kützing	-	1	-	B	i	i	o-α	k
E. silesiacum (Bleisch) Mann (*Cymbella ventricosa* Kütz., *C. ventricosa* var. *hankensis* Skvortzow)	1-6	1-5	2-6	B	i	alf	α	k
Encyonopsis microcephala (Grunow) Krammer (=*Cymbella microcephala* Grunow)	-	+	-	B	i	alf	β	k
Placoneis clementioides (Hustedt) Cox	1	-	-	B	-	alb	-	-

Table 2. (Continued)

Taxa	Areas of Sakhalin Island			Ecological and geographical characteristic				
	East	South	West	B	H	pH	S	G
P. clementis (Grunow) Cox*	-	-	1	B	i	alf	χ-o	b
P. constans (Hustedt) Cox	-	-	1	B	-	-	-	-
P. elginensis (Gregory) Cox	1	1	1	B	i	i	o-β	k
P. gastrum (Ehrenberg) Mereschkowsky (=*Navicula gastrum* Ehrenberg var. *hankensis* Skvortzow)	-	3	-	B	i	i	o-β	k
P. placentula (Ehrenberg) Heinzerling (=*Navicula placentula* (Ehrenberg) Grunow)	-	+	-	B	i	alf	β	k
Family Gomphonemataceae								
Didymosphenia geminata (Lyngbye) M. Schmidt	1-5	1	+	B	i	i	χ	a-a
Gomphoneis olivaceum (Hornemann) Dawson ex Ross et Sims var. *olivaceum* (=*Gomphonema olivaceum* (Hornemann) Brébisson)	1-4	1-3	3-5	B	i	alf	β	b
G. olivaceum var. *calcareum* (Cleve) Hartley (=*Gomphonema olivaceum* var. *calcareum*)	-	+	-	B	i	alf	β	b
G. quadripunctatum (Oestrup) Dawson ex Ross et Sims	1-6	1-2	-	B	i	i	-	b
Gomphonema acuminatum Ehrenberg	1	1	1-5	B	i	alf	β	b
G. affine Kützing (=*G. lanceolatum* Ehrenberg)	2	+	1	B-P	-	-	o-β	k
G. angustatum (Kützing) Rabenhorst	1-3	1	-	B	i	alf	o	b
G. angusticephalum Reichelt et Lange-Bertalot	-	-	1	-	-	-	-	-
G. angustum Agardh (*Gomphonema intricatum* Kützing)	1-3	1-4	2	B	i	alf	o	b
G. augur Ehrenberg var. *augur*	1	1	-	B	i	i	β	k

Taxa	Areas of Sakhalin Island			Ecological and geographical characteristic				
	East	South	West	B	H	pH	S	G
G. augur var. *gautieri* Van Heurck	-	+	+	B	i	i	β	b
G. brebissonii Kützing	1	1-2	-	B	i	alf	β	b
G. clavatum Ehrenberg (=*G. longiceps* Ehrenberg, *G. longiceps* Ehrenberg var. *subclavatum* Grunow)	1	1	+	B	i	i	o	k
G. clevei Fricke	1	1	-	B	-	-	χ	k
G. coronatum Ehrenberg (=*G. acuminatum* Ehrenberg var. *coronata* Ehrenberg)	1	1-3	+	B	i	alf	β	b
G. globiferum Meister	-	1	-	B	-	-	-	-
G. gracile Ehrenberg (=*G. gracile* var. *lanceolatum* Kützing)	-	+	-	B-P	i	alf	β-o	k
G. insigne Gregory*	-	-	1	B	i	-	-	-
G. lagerheimii Cleve	-	1	1-2	B	-	acf	-	a-a
G. micropus Kützing*	-	-	1-2	B	i	alf	o	k
G. minusculum Krasske	-	1	-	-	-	-	-	-
G. minutum (Agardh) Agardh	-	1	2	B	oh	alf	o-β	k
G. parvulum (Kützing) Kützing	1-6	1-5	1-6	B	i	alf	β	b
G. productum (Grunow) Lange-Bertalot et Reichelt (=*G. angustatum* var. *productum* Grunow)	-	+	-	B	i	alf	β	k
G. sarcophagus Gregory (=*G. angustatum* var. *sarcophagus* (Gregory) Grunow)	-	-	+	B	-	-	β-α	-
G. sphaerophorum Ehrenberg	-	+	-	B	i	alf	-	b
G. subtile Ehrenberg	-	+	-	B	i	i	-	b
G. truncatum Ehrenberg (=*G. constrictum* Ehrenberg)	1	1-4	1-3	B	i	alf	β	k
G. truncatum var. *capitatum* (Ehrenberg) Patrick (=G. *constrictum* var. *capitatum* (Ehrenberg) Cleve)	2	1	-	B	i	alf	β	b

Table 2. (Continued)

Taxa	Areas of Sakhalin Island			Ecological and geographical characteristic				
	East	South	West	B	H	pH	S	G
G. ventricosum Gregory	1-2	-	-	B-P	i	i	o-χ	k
G. vibrio Ehrenberg (=*G. intricatum* var. *vibrio* (Ehrenberg) Cleve)	-	1	-	B	-	-	o	-
Reimeria sinuata (Gregory) Kociolek et Stoermer f. *sinuata*	1-3	1-5	1-4	B	i	alf	β	b
R. sinuata f. *antiqua* (Grunow) Kociolek et Stoermer	1	-	1-4	-	-	-	-	-
Order Achnanthales								
Family Achnanthaceae								
Achnanthes brevipes Agardh var. *intermedia* (Kützing) Cleve	1	-	-	B	-	-	-	-
Family Achnanthidiaceae								
Achnanthidium coarctatum Brébisson ex W. Smith (=*Achnanthes coarctata* (Brébisson) Grunow, *A. coarctata* var. *elliptica* Krasske)	1	1	-	B	i	i	o	a-a
A. exiguum (Grunow) Czarnecki (=*Achnanthes exigua* Grunow)	1	1-2	+	B	i	alf	β	k
A. minutissimum (Kützing) Czarnecki (=*Achnanthes minutissima* var. *cryptocephala* Grunow, *A. microcephala* (Kützing) Grunow)	1-6	1-6	3-5	B	i	i	o-β	b
A. pyrenaicum (Hustedt) Kobayasi (=*A. biasolettiana* (Kützing) Grunow)	1	+	-	B	i	-	β-α	k
Achnanthes sp.	-	+	+	-	-	-	-	-
Eucocconeis laevis (Oestrup) Lange-Bertalot	1	1	-	B	-	-	o	-
Karayevia laterostrata (Hustedt) Round et Bukhtiyarova	2	-	-	B	i	i	o	a-a
Lemnicola hungarica (Grunow) Round et Basson (=*Achnanthes hungarica* (Grunow) Grunow, *A. hungarica* Grunow)	1-6	-	-	B	mh	alf	o-α	k
Planothidium conspicuum (A. Mayer) Aboal (=*Achnanthes conspicua* A. Mayer)	1	+	+	B	i	alf	o-α	k

Taxa	Areas of Sakhalin Island			Ecological and geographical characteristic				
	East	South	West	B	H	pH	S	G
P. delicatulum (Kützing) Round et Bukhtiyarova (=*Achnanthes delicatula* (Kützing) Grunow)	-	+	+	B	-	-	-	-
P. ellipticum (Cleve) Edlund (=*Achnanthes lanceolata* var. *elliptica* Cleve)	-	1	+	B	i	alf	-	k
P. hauckianum (Grunow) Round et Bukhtiyarova var. *hauckianum* (=*Achnanthes hauckiana* Grunow)	-	2	-	B	hl	alf	-	k
P. hauckianum var. *rostratum* (Schulz ex Hustedt) Andresen, Stoermer et Kreis (=*Achnanthes hauckiana* var. *rostrata* Schulz)	-	+	-	B	hl	alf	-	b
P. haynaldii (Schaarschmidt) Lange-Bertalot et Genkal (=*Achnanthes lanceolata* var. *haynaldii* (Schaarschmidt) Cleve, *A. lanceolata* f. *capitata* O. Müller, *P. lanceolata* var. *haynaldii* (Schaarschmidt) Bukhtiyarova)	1	1-6	3-6	B	-	alf	β-α	k
P. lanceolatum (Brébisson ex Kützing) Lange-Bertalot (=*Achnanthes lanceolata* (Brébisson) Grunow)	1-2	1-2	2-6	B	i	alf	χ-β	k
P. peragallii (Brun et Heribaud) Round et Bukhtiyarova (=*Achnanthes peragalli* Brun et Heribaud)	1	1	-	B	i	i	o	b
P. rostratum (Oestrup) Round et Bukhtiyarova (=*Achnanthes lanceolata* var. *rostrata* (Oestrup) Hustedt, *A. lanceolata* var. *rostrata* Hustedt)	-	1-3	+	B	i	alf	-	k
Rossithidium linearis (W. Smith) Round et Bukhtiyarova (=*Achnanthes linearis* (W. Smith) Grunow)	1	2-4	-	B	i	i	χ-o	k
R. pusillum (Grunow) Round et Bukhtiyarova (=*Achnanthes pusilla* (Grunow) De Toni)	-	6	-	B	-	-	o	-
Pauliella taeniata (Grunow) Round et Basson (=*Achnanthes taeniata* Grunow)	-	1	-	B	hl	-	-	-
Family Cocconeidaceae								
Cocconeis disculus (Schumann) Cleve	1-2	1	-	B	i	-	-	-

Table 2. (Continued)

Taxa	Areas of Sakhalin Island			Ecological and geographical characteristic				
	East	South	West	B	H	pH	S	G
C. formosa Brun	+	-	-	B	-	-	-	-
C. pediculus Ehrenberg	-	+	+	B	hl	alf	β	k
C. placentula Ehrenberg var. *placentula*	1-4	1-2	1-6	B	i	alf	β	b
C. placentula var. *euglypta* (Ehrenberg) Grunow	1-5	1-6	-	B	i	alf	-	b
C. placentula var. *lineata* (Ehrenberg) Van Heurck	1	1	-	B	i	alf	-	b
C. scutellum Ehrenberg	1	1-6	+	B	hl	-	-	-
Cocconeis sp.	-	+	-	-	-	-	-	-
Order Naviculales								
Family Berkeleyaceae								
Parlibellus crucicula (Smith) Witkowski, Lange-Bertalot et Metzeltin (=*Navicula crucicula* (W. Smith) Donkin)	-	1-3	-	B	mh	i	-	k
P. protracta (Grunow) Witkowski, Lange-Bertalot et Metzeltin	1-2	-	-	B	hl	i	χ-β	k
Family Cavinulaceae								
Cavinula cocconeiformis (Gregory ex Greville) Mann et Stickle	-	1	-	B	i	i	o	a-a
C. lacustris (Gregory) Mann et Strickle (=*Navicula lacustris* Gregory)	-	+	-	B	i	i	o	b
C. pseudoscutiformis (Hustedt) Mann et Stickle	-	2	-	B	i	i	o	a-a
Family Cosmioneidaceae								
Cosmioneis pusilla (W. Smith) Mann et Stickle (=*Navicula pusilla* W.Sm.)	1	1-3	+	B	hl	i	o-β	k
Family Diadesmidaceae								

Taxa	Areas of Sakhalin Island			Ecological and geographical characteristic				
	East	South	West	B	H	pH	S	G
Luticola mutica (Kützing) Mann (=*Navicula mutica* Kützing, *Navicula mutica* var. *binodis* Hustedt)	1-5	1	+	B	i	i	o-β	a-a
L. nivalis (Ehrenberg) Mann*	-	-	1	B	hl	i	-	k
L. ventricosa (Kützing) Mann (=*Navicula mutica* var. *ventricosa* (Kützing) Cleve)	-	-	+	B-P	hl	alf	χ-o	k
Family Amphipleuraceae								
Amphipleura pellucida (Kützing) Kützing	1-3	-	3	B	i	alf	α-β	k
Frustulia ampipleuroides (Grunow) Cleve-Euler	1	-	3	B	hb	acf	-	a-a
F. rhomboides (Ehrenberg) De Toni f. *rhomboides*	1-2	1	1	B	hb	acf	o-χ	a-a
F. rhomboids f. *undulata* Hustedt	-	1	-	B	-	-	-	-
F. vulgaris (Thwaites) De Toni	1-2	1	+	B	hb	alf	o	b
Family Brachysiraceae								
Brachysira serians (Brébisson ex Kützing) Round et Mann	-	1	-	B	oh	acb	χ-o	k
Family Neidiaceae								
Neidium affine (Ehrenberg) Pfitzer var. *affine* Cleve (=*N. affine* var. *amphirhynchus* (Ehrenberg) Cleve)	-	1	-	B	i	alf	o	b
N. affine (Ehrenberg) Pfitzer var. *medium* Cleve (=*N. affine* var. *affine* f. *medium* Cleve)	-	1	-	B	i	alf	o	b
N. ampliatum (Ehrenberg) Krammer	1	1	1	B	hb	i	o	k
N. bisulcatum (Lagerstadt) Cleve	1	1	-	B	hb	i	o-β	b
N. dubium (Ehrenberg) Cleve	1	1	-	B	i	alf	χ	k
N. iridis (Ehrenberg) Cleve	1	1	1	B	hb	i	o- χ	k
N. productum (W. Smith) Cleve	1	+	+	B	i	acf	o-β	k

Table 2. (Continued)

Taxa	Areas of Sakhalin Island			Ecological and geographical characteristic				
	East	South	West	B	H	pH	S	G
Neidium sp.	1	-	1	-	-	-	-	-
Family Sellaphoraceae								
Fallacia pygmaea (Kützing) Stickle et Mann (=*Navicula pygmaea* Kützing)	1	1	+	B	mh	alf	α	b
Sellaphora bacillum (Ehrenberg) Mann (=*Navicula bacillum* Ehrenberg)	1	1	-	B	i	alf	χ-o	k
S. laevissima (Kützing) Mann	-	-	1	B	-	-	-	-
S. pupula (Kützing) Mann	1-2	1	-	B	hl	i	o-χ	k
S. pupula f. *rostrata* (Hustedt) Bukhtiyarova (=*Navicula pupula* var. *rostrata* Hustedt)	-	+	-	B	hl	i	β	k
S. rectangularis (Gregory) Lange-Bertalot et Metzeltin (=*Navicula pupula* var. *rectangularis* (Gregory) Grunow)	-	-	+	B	hl	i	-	k
S. seminulum (Grunow) Mann (=*Navicula seminulum* Grunow)	-	+	-	B-P	i	i	χ-o	k
S. wummensis Johansen (=*Navicula pupula* var. *elliptica* Hustedt)	-	+	-	B	hl	i	-	k
Family Pinnulariaceae								
Pinnularia acidojaponica Idei et Kobayasi	1	-	-	B	-	acf	-	-
P. acrosphaeria W. Smith	-	1	-	B	i	alf	-	k
P. acuminata W. Smith	-	-	1	B	i	-	o	k
P. alpina W. Smith	1-2	1	-	B	i	-	-	a-a
P. angulosa Krammer	1	-	-	B	-	-	-	-
P. angusta (Cleve) Krammer*	-	-	1	B	-	acf	-	-
P. biceps Gregory	1	1-2	-	B	i	acf	β-o	k

Taxa	Areas of Sakhalin Island			Ecological and geographical characteristic				
	East	South	West	B	H	pH	S	G
P. borealis Ehrenberg	1-5	1	1	B	i	i	χ	a-a
P. brebissonii (Kützing) Rabenhorst	-	2	-	B	i	i	α-β	b
P. crucifera Cleve-Euler (=*P. brevicostata* Cleve)	-	+	-	B	i	i	-	k
P. dactylus Ehrenberg	-	+	+	B	oh	acf	-	b
P. decrescens (Grunow) Krammer	-	-	1	B	-	-	-	-
P. divergens W. Smith var. *media* Krammer	1	-	-	B	-	-	-	-
P. eifelana Krammer	1	-	-	B	-	-	-	-
P. gibba Ehrenberg	+	1	+	B	i	i	o-β	b
P. globiceps Gregory	1	-	-	B	i	i	-	b
P. grunowii Krammer (=*P. interrupta* W. Smith, *P. mesolepta* (Ehrenberg) W. Smith)	1	1	1-2	B	-	-	-	-
P. inconstans A. Mayer	-	-	1	B	-	-	-	-
P. intermedia (Lagerstedt) Cleve*	-	-	1	B	i	i	χ	b
P. karelica Cleve	1	-	-	B	i	i	o	a-a
P. lata (Brébisson) Rabenhorst	-	-	-	B	i	acf	o	b
P. macilenta (Ehrenberg) Cleve	1	-	1	B	-	-	o	b
P. microstauron (Ehrenberg) Cleve	1	-	+	B	i	i	o	b
P. neomajor Krammer (=P. major (Kützing) Cleve	1	+	-	B	i	i	χ	k
P. nobilis (Ehrenberg) Ehrenberg	-	-	3	B	i	acf	χ	b
P. nodosa (Ehrenberg) W. Smith	1	1	-	B	i	i	o	a-a
P. nodosa var. *percapitata* Krammer	-	1	-	B	-	-	-	-
P. obscura Krasske	1	-	-	B	-	-	-	-

Table 2. (Continued)

Taxa	Areas of Sakhalin Island			Ecological and geographical characteristic				
	East	South	West	B	H	pH	S	G
P. parvulissima Krammer	1	-	-	B	-	-	-	-
P. pseudogibba Krammer	-	1	-	B	-	-	-	-
P. rhombarea Krammer	1	-	-	B	-	-	-	-
P. rupestris Hantzsch	1	1	-	B	-	acf	-	-
P. septentrionales Krammer	1	-	-	B	-	-	-	-
P. subbrevistriata Krammer	-	-	1	-	-	-	-	-
P. subcapitata Gregory	1	-	-	B	i	i	χ-o	k
P. subgibba Krammer var. *subgibba* (=*P. stauroptera* Grunow, *P. stauroptera* var. *interrupta* Cleve)	1	+	1	B	-	-	o	-
P. subgibba var. *undulata* Krammer*	-	-	1	B	-	-	o	-
P. subinterrupta Krammer et Schroeter (=*P. interrupta* f. *minutissima* Hustedt)	-	+	-	B	-	-	-	-
P. subsolaris (Grunow) Cleve	-	-	+	B	oh	i	-	a-a
P. subundulata Østrup	1	-	-	-	-	-	-	-
P. viridiformis Krammer	1	1	1	B	-	-	o	-
P. viridis (Nitzsch) Ehrenberg	-	1	+	B	i	i	β	b
Pinnularia sp.	1	+	+	-	-	-	-	-
Pinnunavis elegans (W. Smith) Okuno (=*Navicula elegans* W. Smith)	1	-	-	-	hl	-	-	-
Family Diploneidaceae								
Decussata placenta (Ehrenberg) Lange-Bertalot et Metzeltin (=*Navicula placenta* Ehrenberg)	-	1	-	B	hb	acf	o	k

Taxa	Areas of Sakhalin Island			Ecological and geographical characteristic				
	East	South	West	B	H	pH	S	G
Diploneis elliptica (Kützing) Cleve	1	1	-	B	i	alf	o	k
D. finnica (Ehrenberg) Cleve	-	1-2	-	B	i	i	-	a-a
D. interrupta (Kützing) Cleve	1-2	1	-	B	mh	i	-	k
D. oblongella (Nägeli) Ross	1	1	-	B	i	alf	o-α	k
D. ovalis (Hilse) Cleve	1	1-2	1	B	hl	alf	β	b
D. parma Cleve	-	+	1	B	i	alf	o-β	-
D. pseudovalis Hustedt	1	-	-	B	mh	alf	-	b
D. smithii (Brébisson) Cleve var. *smithii*	+	1	-	B	mh	alf	-	k
D. smithii var. *rhombica* Mereschkowsky	1	-	-	B	hl	-	-	-
Family Naviculaceae								
Caloneis bacillum (Grunow) Cleve*	-	-	1	B-P	i	alf	o	k
C. brevis Greville	-	1	-	-	-	-	-	-
C. latiuscula (Kützing) Cleve	-	1	-	B	i	i	o	k
C. molaris (Grunow) Krammer*	-	-	1	B	i	i	-	a-a
C. schumanniana (Grunow) Cleve (=*C. limosa* (Kützing) Patrick)	-	1	-	B-P	i	alf	o-χ	k
C. silicula (Ehrenberg) Cleve	1	1-2	1-2	B	i	alb	o	k
Caloneis sp.	1-3	+	-	B	-	-	-	-
Chamaepinnularia krookii (Grunow) Lange-Bertalot & Krammer	1	1	1	B	-	-	-	-
Geissleria decussis (Oestrup) Lange-Bertalot et Metzeltin (=*Navicula decussis* Oestrup)	-	1	-	B	-	-	β-o	-
Haslea spicula (Hickie) Bukhtiyarova (=*Navicula spicula* (Hickie) Cleve)	-	1-4	-	B-P	mh	-	-	k
Hippodonta capitata (Ehrenberg) Lange-Bertalot, Metzeltin et Witkowski	1-2	+	+	B	hl	alf	χ-o	k

Table 2. (Continued)

Taxa	Areas of Sakhalin Island			Ecological and geographical characteristic				
	East	South	West	B	H	pH	S	G
H. hungarica (Grunow) Lange-Bertalot, Metzeltin et Witkowski	1	+	-	B	hl	alf	β	b
Navicula avenacea (Brébisson et Godey) Brébisson ex Grunow	1-5	1	2-6	B	i	acf	β	-
N. brasiliensis Grunow	-	+	-	-	-	-	-	-
N. capitatoradiata Germain	-	1	+	B-P	i	-	β	k
N. cari Ehrenberg	-	+	-	B-P	i	i	β-α	k
N. cincta (Ehrenberg) Ralfs	-	4-5	-	B	hl	alf	χ-o	k
N. concentrica Carter	1	-	-	B	-	-	χ	-
N. cryptocephala Kützing	1-5	1-3	3-4	B-P	hl	alf	α	k
N. cryptotenella Lange-Bertalot	1-5	2-4	+	B	i	alf	β	k
N. digitoradiata (Gregory) Ralfs	1-4	-	-	B	hl	alf	-	k
N. directa W. Smith (=*N. directa* (W. Smith) Ralfs)	1	-	-	B-P	mh	-	-	-
N. gelida Grunow var. *subimpressa* (Grunow) Cleve	-	+	-	-	-	-	-	-
N. integra (W. Smith) Ralfs	1	1	-	B	mh	i	χ-o	a-a
N. interglacialis Hustedt (=*N. grevillei* Agardh)	-	+	-	B	i	i	-	b
N. jentzschii Grunow	-	1	-	B	i	i	-	b
N. lanceolata (Agardh) Kützing	+	+	+	B	i	alf	χ-β	k
N. menisculus Schumann	1	1	-	B	hl	alf	β-α	k
N. meniscus Schumann (=*N. menisculus* var. *meniscus* (Schumann) Hustedt)	-	+	+	B	hl	alf	-	k
N. peregrina (Ehrenberg) Kützing var. *peregrina*	1	+	-	B	mh	alf	-	k
N. peregrina var. *kefvingensis* (Ehrenberg) Cleve	-	3	-	B	mh	-	-	k

Taxa	Areas of Sakhalin Island			Ecological and geographical characteristic				
	East	South	West	B	H	pH	S	G
N. phyllepta Kützing	-	3-6	-	B	hl	-	-	k
N. radiosa Kützing	1-3	1-4	2-4	B	i	i	o-β	k
N. rhynchocephala Kützing	1-3	1-4	1	B	hl	-	β-α	-
N. salinarum Grunow	-	-	-	B	mh	-	-	-
N. salinicola Hustedt (=*N. incerta* Grunow)	-	-	+	B	hl	-	-	-
N. slesvicensis Grunow	1-4	1-2	1	B	hl	i	β	k
N. transitans Cleve	-	+	-	-	-	-	-	-
N. tripunctata (O. Müll.) Bory (=*N. gracilis* Ehrenberg)	1	1-3	-	B	i	i	β	k
N. viridula (Kützing) Ehrenberg (=*N. viridula* Kützing)	2	+	-	B	hl	alf	o	k
N. vulpina Kützing	-	+	+	B	i	alf	o	b
Navicula sp. 1	-	+	-	-	-	-	-	-
Navicula sp. 2	-	+	-	-	-	-	-	-
Navicula sp. 3	-	+	-	-	-	-	-	-
Naviculadicta laterostrata Hustedt	-	+	-	B	i	-	-	-
Trachyneis aspera (Ehrenberg) Cleve	1	-	-	-	-	-	-	-
Family Plagiotropidaceae								
Plagiotropis maxima (Gregory) O. Kuntze (=*Tropidoneis maxima* var. *dubia* (Cleve & Grunow) Cleve)	-	+	-	B	-	-	-	-
Family Pleurosigmataceae								
Gyrosigma acuminatum (Kützing) Rabenhorst var. *acuminatum*	1	1-4	1	B	i	alb	β	b
G. acuminatum var. *gallica* Grunow	-	1	-	B	hl	-	-	k
G. attenuatum (Kützing) Rabenhorst	1	+	-	B-P	i	alf	χ	k

Table 2. (Continued)

Taxa	Areas of Sakhalin Island			Ecological and geographical characteristic				
	East	South	West	B	H	pH	S	G
G. baikalensis Skvortzow	-	-	+	B	i	-	-	-
G. fasciola (Ehrenberg) Griffith et Henfrey (= *G. fasciola* Ehrenberg)	-	+	-	B	mh	-	-	-
G. spenserii (Quekett) Griffith et Henfrey (= *G. kützingii* (Grunow) Cleve)	-	+	-	B	mh	alf	o	k
Pleurosigma angulatum (Quekett) W. Smith	-	+	-	B	hl	-	-	-
P. longum Cleve	-	1	-	B	-	-	-	-
Pleurosigma sp.	-	+	-	-	-	-	-	-
Family Stauroneidaceae								
Craticula ambigua (Ehrenberg) Mann	-	-	1	B	i	alf	o	k
C. cuspidata (Kützing) Mann	1	-	-	B	i	alf	o-β	k
C. halophila (Grunow) Mann (=*Navicula simplex* Krasske)	-	+	-	B	mh	alf	-	k
Stauroneis anceps Ehrenberg var. *anceps*	1-4	1	1-3	B	i	i	χ	k
S. anceps var. *gracilis* (Ehrenberg) Cleve	1	-	-	B	-	-	o-χ	-
S. legumen (Ehrenberg) Kützing	-	+	-	B	i	i	-	-
S. nobilis Schumann f. *alabamae* (Heiden) Cleve-Euler (=*S. alabamae* Heiden)	+	-	-	B	-	acf	-	-
S. phoenicenteron (Nitzsch) Ehrenberg (=*S. phoenicenteron* Ehrenberg)	1	1	1	B	i	i	χ-o	k
S. smithii Grunow	-	1-2	-	B-P	i	alf	χ-o	k
Order Thalassiophysales								
Family Catenulaceae								
Amphora coffeaeformis (Agardh) Kützing (=*A. coffeaeformis* Ag.)	-	4	-	B	mh	alf	-	k
A. commutata Grunow	-	1-4	-	B	mh	-	-	k

Taxa	Areas of Sakhalin Island			Ecological and geographical characteristic				
	East	South	West	B	H	pH	S	G
A. holsatica Hustedt	1	-	-	P	mh	-	-	k
A. libyca Ehrenberg (=*A. copulata* (Kützing) Schoeman et Archibald	1-2	1	1	B	i	alf	-	k
A. normanii Rabenhorst	1	-	-	B	hb	alf	β-α	b
A. ocellata Donkin	-	+	-	B	-	-	-	-
A. ovalis (Kützing) Kützing	1	+	+	B	i	alb	o-β	k
A. pediculus (Kützing) Grunow (=*A. perpusilla* Grunow)	1	1	-	B	i	alb	β	k
A. veneta Kützing	1	3-5	-	B	i	alf	o	k
Amphora sp.	-	+	+	-	-	-	-	-
Order Bacillariales								
Family Bacillariaceae								
Bacillaria paxillifer (O. Müller) Hendey (=*B. paradoxa* Gmelin)	-	1-5	-	P	mh	alb	β	k
B. socialis Grunov	1	-	-	B-P	mh	-	-	-
Cylindrotheca closterium (Ehrenberg) Lewin et Reimann (=*C. closterium* (Ehrenberg) Reimer et Lewin)	-	+	-	-	-	-	-	-
Hantzschia amphioxys (Ehrenberg) Grunow	1	1	1-4	B	i	alf	α	k
Hantzschia sp.	-	+	-	-	-	-	-	-
Nitzschia acicularis (Kützing) W. Smith	1-6	1	+	B-P	i	alf	o-β	k
N. amphibia Grunow	1-2	1	-	B-P	i	alf	o	k
N. brevissima Grunow (=*N. parvula* Lewis)	+	+	1	B	hl	i	o-β	k
N. capitellata Hustedt var. *capitellata**	-	-	2	B	i	alb	o	k

Table 2. (Continued)

Taxa	Areas of Sakhalin Island			Ecological and geographical characteristic				
	East	South	West	B	H	pH	S	G
N. capitellata var. *tenuirostris* (Grunow) Bukhtiyarova (=*N. palea* var. *tenuirostris* Grunow)	-	+	-	B	-	-	-	-
N. clausii Hantzsch	+	1	-	B	mh	acf	o-α	k
N. communis Rabenhorst	1	-	-	B-P	i	alf	o	k
N. commutatoides Lange-Bertalot	1	-	-	-	hl	-	-	-
N. compressa (Bailey) Boyer var. *elongata* (Grunow) Lange-Bertalot	1	-	-	B	hl	-	-	k
N. dissipata (Kützing) Grunow	1-4	1-2	2	B	i	alf	o-β	b
N. dubia W. Smith	-	-	1	B-P	hl	-	o-β	k
N. filiformis (W. Smith) Van Heurck*	-	-	1	B	hl	-	χ	k
N. flexa Schumann	1	-	-	B	oh	alf	o-α	-
N. fonticola Grunow	1-2	1-6	2	B	i	alf	o-β	b
N. frigida Grunow	-	+	-	-	-	-	-	-
N. frustulum (Kützing) Grunow	1	1	+	B	hl	alb	o	k
N. gracilis Hantzsch (=*N. gracilis* var. *capitata* Wislouch et Poretzky)	-	3	-	B	i	i	β	b
N. heufleriana Grunow	1	-	-	B	i	alf	o-β	k
N. lanceola Grunow	-	1	-	B	-	-	-	-
N. lanceolata W. Smith	-	+	-	-	hl	-	-	-
N. linearis (Agardh) W. Smith	1-2	1	1-4	B	i	i	χ	b
N. lorenziana Grunow	1	-	-	B	mh	-	-	k
N. microcephala Grunow	-	+	-	B	hl	alb	o-β	k

Taxa	Areas of Sakhalin Island			Ecological and geographical characteristic				
	East	South	West	B	H	pH	S	G
N. nana Grunow	1-3	-	-	B	mh	-	-	b
N. obtusa W. Smith	+	+	-	B	mh	-	β	k
N. palea (Kützing) W. Smith var. *palea*	1-6	1-6	+	B	i	i	α	k
N. palea var. *capitata* Wislouch et Poretzky*	-	-	2	B	i	i	β	k
N. paleacea (Grunow) Grunow (=*N. holsatica* Grunow, *N. holsatica* Hustedt)	1-3	1-6	3	P	i	alf	β	k
N. recta Hantzsch	-	+	-	B	i	alf	α-β	b
N. reversa W. Smith (=*N. longissima* (Brébisson) Ralfs)	1	+	-	P	hl	-	-	k
N. sigma (Kützing) W. Smith var. *sigma*	1	1-3	2	B	-	-	-	k
N. sigma (Kützing) W. Smith var. *curvula* (Ehrenberg) Brun	-	2	-	B	mh	alf	-	k
N. sublinearis Hustedt	-	+	+	B-P	i	i	o-β	k
N. vermicularis (Kützing) Hantzsch	1	3	-	B	i	alf	β	k
Nitzschia sp.	-	+	+	-	-	-	-	-
Pseudo-nitzschia pungens (Grunow ex Cleve) Hasle	-	+	-	-	-	-	-	-
Pseudo-nitzschia sp.	-	+	-	B	mh	alf	β	k
Tryblionella apiculata Gregory	1-5	1-5	-	B	mh	alf	β	k
T. debilis Arnott (=*Nitzschia tryblionella* var. *debilis* (Arnott) Grunow)	-	+	-	B-P	hl	alf	o	k
T. gracilis W. Smith (=*Nitzschia tryblionella* Grunow, *N. tryblionella* Hantzsch)	-	+	-	B	hl	alf	o	k
T. granulata (Grunow) Mann	1	-	-	B-P	-	-	-	-
T. hungarica (Grunow) Mann	1	-	-	B-P	mh	alf	α-β	k

Table 2. (Continued)

Taxa	Areas of Sakhalin Island			Ecological and geographical characteristic				
	East	South	West	B	H	pH	S	G
T. levidensis (W. Smith) Grunow (=*Nitzschia tryblionella* var. *levidensis* (W. Smith) Grunow, *N. levidensis* (W. Smith) Grunow)	1-2	1-5	1-2	B	hl	alf	α	b
T. marginulata (Grunow) Mann	-	3	-	-	hl	-	-	-
T. punctata W. Smith	1	-	-	B	mh	-	-	k
T. victoriae Grunow (=*Nitzschia tryblionella* var. *victoriae* Grunow)	-	1-4	-	B	hl	-	-	b
Order Rhopalodiales								
Family Rhopalodiaceae								
Epithemia adnata (Kützing) Brébisson (=*E. zebra* (Ehrenberg) Kützing)	1-2	1-3	1	B	i	alb	β-α	k
E. adnata var. porcellus (Kützing) Ross (=*E. zebra* var. *porcellus* (Kützing) Grunow)	1	1-2	-	B	i	alb	β	k
E. argus (Ehrenberg) Kützing var. *argus* (=*E. argus* Kützing)	-	+	+	B-P	i	i	-	k
E. argus var. *alpestris* (W. Smith) Grunow (=*E. argus* var. *capitata* Fricke)	-	+	-	B	i	i	-	b
E. sorex Kützing	1	-	-	B	i	alf	o-α	k
E. turgida (Ehrenberg) Kützing	1	-	1-4	B	i	alf	o	k
Epithemia sp.	-	-	+	-	-	-	-	-
Rhopalodia acuminata Krammer	1	-	-	B	hl	-	-	-
Rh. gibba (Ehrenberg) O. Müller var. *gibba*	1-2	1-3	1-5	B	i	alb	χ-o	k
Rh. gibba var. *parallela* (Grunow) H. et M. Peragallo	1	1-3	-	B	i	alf	o	a-a
Rh. gibberula (Ehrenberg) O.F. Müller	1	+	-	B	mh	i	-	k
Rh. musculus (Kützing) O. Müller	1	1	-	B-P	mh	alb	-	k

Taxa	Areas of Sakhalin Island			Ecological and geographical characteristic				
	East	South	West	B	H	pH	S	G
Rh. operculata (Agardh) Hakansson	1	-	-	B	-	-	-	-
Rh. rupestris (W. Smith) Krammer	1	-	-	B	-	-	o	a-a
Rhopalodia sp.	-	+	-	-	-	-	-	-
Order Surirellales								
Family Entomoneidaceae								
Entomoneis alata (Ehrenberg) Ehrenberg (=*Amphiprora alata* Kützing)	-	1-2	-	B-P	mh	alf	-	k
E. ornata (Bailey) Reimer (=*Amphiprora ornata* Bailey)	-	+	+	B	-	-	o-χ	k
E. paludosa (W. Smith) Reimer var (=*Amphiprora paludosa* W. Smith)	-	+	-	B	-	-	-	k
Family Surirellaceae								
Campylodiscus echeneis Ehrenberg	-	3	-	P	hl	-	-	k
C. hibernicus Ehrenberg	-	1	-	B	i	i	o	b
Campylodiscus sp.	-	+	-	-	-	-	-	-
Cymatopleura solea (Brébisson) W. Smith	1	1	1-2	B	i	alf	β-α	k
Petrodictyon gemma (Ehrenberg) Mann (=*Surirella gemma* Ehrenberg)	-	+	-	B	hl	-	-	-
Surirella angusta Kützing (=*S. angustata* Kützing)	1-3	1-2	1	B	-	-	o	a-a
S. biseriata Brébisson (=*S. biseriata* var. *constricta* Grunow)	1	-	+	B-P	i	alf	o-β	k
S. brebissonii Krammer & Lange-Bertalot var. *brebissonii*	1-2	1-2	-	B	i	i	β	k
S. brebissonii var. *kuetzingii* Krammer et Lange-Bertalot (=*S. ovata* Kützing)	1-2	1	1-4	B	-	-	β-α	-
S. capronii Brébisson	-	+	+	B-P	i	i	χ	k
S. elegans Ehrenberg	1	-	2	B-P	i	alb	o	k
S. fastuosa Ehrenberg	-	1	-	-	-	-	-	-

Table 2. (Continued)

Taxa	Areas of Sakhalin Island			Ecological and geographical characteristic				
	East	South	West	B	H	pH	S	G
S. gracilis (W. Smith) Grunow	1	1	-	B	i	-	-	a-a
S. linearis W. Smith (=*S. linearis* var. *constricta* (Ehrenberg) Grunow)	1	+	+	B-P	i	i	β	a-a
S. minuta Brébisson (=*S. ovata* var. *salina* (W. Smith) Hustedt)	1-2	1	+	B	i	alf	-	b
S. ovalis Brébisson	-	+	+	B-P	mh	alf	o	k
S. robusta Ehrenberg	-	1	-	B-P	hb	i	β-o	k
S. splendida (Ehrenberg) Kützing	1	1	-	B-P	i	alf	o-β	k
S. tenera Gregory	1	+	+	B-P	i	alf	o	k
S. tenera var. *nervosa* Schmidt (=*S. tenera* var. *nervosa* A.S.)	-	1-3	-	B-P	i	alf	β-α	k
S. tenuis A. Mayer	-	1	-	-	-	-	-	-
S. tientsinensis Skvortzow	1	-	-	B	mh	-	-	b
S. ussuriensis Skvortzow	-	+	-	-	-	-	-	-
Surirella sp.	-	+	+	-	-	-	-	-

Notes: East – streams flowing into the Okhotsk Sea, South – the Japan Sea, West – the Amur Estuary and the straits of Tartary. To estimate the frequency of taxa occurrence at the stations, we used the six point scale: 1 – solitary (1-5 cells in the slide); 2 – rare (10-15 cells in the slide); 3 – not infrequent (25-30 cells in the slide); 4 – frequent (1 cell in each row of the cover glass at magnification with immersion); 5 – very frequent (several cells under the same conditions); 6 – in bulk (several cells in each visual field under the same conditions) (Korde, 1956); «+» – frequency of taxa occurrence is not known. B (biotope): P – planktonic, B-P – benthic-planktonic, B – benthic, E – epiphytic, B-E – benthic-epiphytic. H (relation to salinity): eu – euhalobic, mh – mesohalobic, hl – halophilous, hb – halophobic, i – indifferent. pH (relation to pH of water): alf – alkaliphilous, alb – alkalibiontic, acf – acidophilous, i – indifferent. S (relation to saprobity of water): χ – xenosaprobous, χ-o – xeno-oligosaprobous, o-χ – oligo-xenosaprobous, χ-β – xeno-betamesosaprobous, o – oligosaprobous, o-β – oligo-betamesosaprobous, β – betamesosaprobous, o-α – oligo-alphamesosaprobous, β-α – beta-alphamesosaprobous, α-β – alpha-betamesosaprobous, α – alphamesosaprobous, ρ - polysaprobous. «-» – data absent. G (geographical distribution): a-a – arctic-alpine, b – boreal, k – cosmopolitan

Table 3. Diatom algae of the Sakhalin Island arranged by ecological groups and types of geographical distribution

Ecological groups and types of geographical distribution	Sakhalin	
	Number of taxa	Percent (%)
Biotope		
benthic	136	64.5
planktonic	42	8.0
benthic-planktonic	64	12.3
epiphythic	2	0.4
benthic-epiphythic	1	0.2
N/A	76	14.6
Total	521	100
Relation to water salinity		
mesohalobic	47	9.0
halophilic	62	11.9
indifferent	211	40.5
halophobic	32	6.1
N/A	169	32.5
Total	521	100
Relation to water pH		
alkalibiontic	21	4.0
alkaliphilous	149	28.6
indifferent	90	17.3
acidobiontic	1	0.2
acidophilous	47	9.0
N/A	213	40.9
Total	521	100
Relation to water saprobity		
ksenosaprobous (χ, χ-o)	42	8.1
oligosaprobous (o-χ, χ-β, o, o-β)	126	24.2
betamesosaprobous (β-o, o-α, β, β-α)	93	17.8
alphamesosaprobous (α-β, β-ρ, α, α-ρ)	16	3.1
polysaprobous (ρ-α, ρ)	1	0.2
N/A	243	46.6
Total	521	100

                    T. V. Nikulina

Table 3 (Continued).

Ecological groups and types of geographical distribution	Sakhalin	
	Number of taxa	Percent (%)
Geographical distribution		
cosmopolitan	209	40.1
boreal	81	15.6
arctic-alpine	44	8.4
N/A	187	35.9
Total	521	100

Assessment of water quality was conducted by Pantle-Buck's method as modified by Sládeček (Pantle, Buck, 1955; Sládeček, 1967) using the presence of a certain algal species as indicator of organic pollution. Index of water saprobity (S) of the eastern part of Sakhalin changes from 1.12 to 1.46, showing oligosaprobous zone, II class of water quality. Index of water saprobity of the southern part part of the island S=1.38–1.46, showing oligosaprobous– betamesosaprobous zone, II-III class of water quality. Index of water saprobity of the western part changes from 0.5 to 2.12, showing ksenosaprobous– oligosaprobous zone, I–II class of water quality (Table 4). This means that the rivers and lakes of the Sakhalin Island are classified as very clean with low organic pollution.

Table 4. Water saprobity data for main rivers and reservoirs of the Sakhalin Island

Water body	Index of saprobity (S)	Degree of saprobity	Zone of saprobity	Class of quality
East of Sakhalin Island				
Tym River, Poronay River	1.22-1.46	o–o-β	oligosaprobous	II
Streams	1.12-1.33	o		
South of Sakhalin Island				
Lakes	1.43-2.44	o-β–β-α	oligosaprobous–	II-III
Streams	1.38-2.25	o–β	betamesosaprobous	
West of Sakhalin Island				
Lake Sladkoe and small lakes	0.5-2.12	χ-o–β	ksenosaprobous– oligosaprobous	I-II
Streams	1.07-1.19	o	oligosaprobous	II

Conclusion

The diatom flora of fresh and blackish water bodies of the Sakhalin Island includes 489 species (521 species, varieties, and forms) from three classes: Coscinodiscophyceae, Fragilariophyceae, and Bacillariophyceae. More than 60 taxa of diatoms are dominant in the algal communities of water bodies of Sakhalin. Benthic organisms, indifferent to water salinity, alkaliphilous, and cosmopolitan species are prevailed in the diatom flora. Rivers and lakes of the Sakhalin Island are classified as very clean with low organic pollution.

Acknowledgments

I am very thankful to T. I. Arefina-Armitage, Dr. E. M. Sayenko (Institute of Biology and Soil Sciences FEB RAS), E. Yu. Matveenko for help in preparation the text of the manuscript. The work was supported by Grant of the Russian Academy of Science JBS RAN (ОБН РАН) №12-1-П30-01.

References

Barinova, S. S., Medvedeva, L. A. & Anissimova, O. V. (2006). *Diversity of algal indicators in environmental assessment.* Tel-Aviv: Piles Studio. 498 p. (In Russian).

Bukhtiyarova, L. N. (1999). *Diatoms of Ukraine. Inland waters.* Kiev. 133 p.

Chudaeva, V. A. (2002). *Chemical elements migration in the waters of the Far East.* Vladivostok: Dalnauka. 392 p. (In Russian).

Genkal, S. I., Motylkova, I. V. & Konovalova, N. V. (2011). New data on the flora of diatom algae (Centrophyceae) in waterbodies of Sakhalin Island. *Inland Water Biology*, v. *4* (4). 408–418.

Guiry, M. D. & Guiry, G. M. (2013). *AlgaeBase.* World-wide electronic publication, National University of Ireland, Galway (available on internet at http://www.algaebase.org) searched on 08 July 2013.

Hammer, Ø., Harper, D. A. T. & Ryan, P. D. (2007). PAST – PAlaeontological STatistics, version 1.89. World Wide Web electronic publication (available on internet at http://folk.uio.no/ohammer/past/) on 30 January 2013.

Hartley, B., Barber, H. G. & Carter, J. R. (1996). *An atlas of British diatoms.* England: Biopress Ltd. 601 p.

Kalganova, T. N & Gertsog, Ya. V. (2012). Comparative characteristics of phytoplankton from some lakes from south Sakhalin in autumn 2004. In A. A. Vasilevskyi (Ed.), *Proceeding of Materials of XXXIV Scientific Conference lecturers, postgraduates and members of SakhGU* (v. *VI*, 38–50). Uzhno-Sakhalinsk: Publishing house of SakhGU. (In Russian).

Karpunin, A. M., Mamonov, S. V., Mironenko, O. A. & Sokolov, A. R. (1998). *Geological landmarks of nature in Russia.* St-Petersburg: Lorien, 200 p. (in Russian).

Knjazev, V. N. & Kalganova, T. N. (2000a). The evolution of phytoplankton of lakes from northwest Sakhalin in the summer and autumn of 1993–1994. In L. I. Rubleva (Ed.), *Proceeding of Materials of XXXIV Scientific Conference lecturer of SakhGU* (p. VI, 29–36). Uzhno-Sakhalinsk: Publishing house of SakhGU. (In Russian).

Knjazev, V. N. & Kalganova, T. N. (2000b). Phytoplankton productivity of the lake Sladkoe (Sakhalin Island). In L. I. Rubleva (Ed.), *Proceeding of Materials of XXXIV Scientific Conference lecturer of SakhGU* (p. VI, 36–41). Uzhno-Sakhalinsk: Publishing house of SakhGU. (In Russian).

Konovalova, N. V. & Motylkova, I. V. (2006). The phytoplankton of Tunaicha Lake (Southern Sakhalin). *Proceedings of the 21st International Symposium on Okhotsk Sea & Sea Ice* (200–204). Mombetsu, Hokkaido, Japan: The Okhotsk Sea and Cold Ocean Research Association.

Konovalova, N. V. & Motylkova, I. V. (2008). Microperiphyton of the Poronai River (Sakhalin Island). In S. G. Inge-Vechtomov & A. F. Alimov (Eds.), *Abstract book of International scientific-and-practical Conference "Periphyton and fouling: Theory and practice"*, (50–51). St-Petersburg: Russian Academy of Science. (In Russian).

Konovalova, N. V. & Motylkova, I. V. (2011). Phytoperiphyton the lower part of the Tym River in September, 2009 (Sakhalin Island). *Proceedings of the II International Scientific Conference "The diversity of soil and biota in North and Central Asia"*, (v. *2*, p. 194). Ulan-Ude: Publishing house of BSC SD RAN. (In Russian).

Koptyaeva, T. F. (1964). The phytoplankton of Vavai Lake from South Sakhalin. In O. A. Klyuchareva (Ed.), *Lakes of South Sakhalin and fish fauna*, (141–153). Moscow: Publishing house of Moscow State University. (In Russian).

Korde, N. V. (1956). Survey procedure of biological study of bottom sediments. In E. N. Pavlovsky & V. I. Zhadin (Eds.), *The life of fresh waters of the USSR*, (v. *4*, 383–413). Moscow, Leningrad: Publishing house of Academy Sciences of the USSR. (In Russian).

Krammer, K. (2000). *Diatoms of Europe. Diatoms of the European inland waters and comparable habitats: The genus Pinnularia*, (V. *1*). Ruggell: A.R.G. Gantner Verlag K.G. 703 p.

Krammer, K. (2002). *Diatoms of Europe. Diatoms of the European inland waters and comparable habitats: Cymbella*, (V. *3*). Ruggell: A.R.G. Ganter Verlag K.G. 584 p.

Krammer, K. (2003). *Diatoms of Europe. Diatoms of the European inland waters and comparable habitats: Cymbopleura, Delicata, Navicymbula, Gomphocymbellopsis, Afrocymbella*, (V. *4*). Ruggell: A.R.G. Ganter Verlag K.G. 530 p.

Krammer, K. & Lange-Bertalot, H. (1986). *Süßwasserflora von Mitteleuropa: Bacillariophyceae: Naviculaceae* (Bd. 2/1). Jena: Gustav Fischer Verlag. 860 p.

Krammer, K. & Lange-Bertalot, H. (1988). *Süßwasserflora von Mitteleuropa: Bacillariophyceae: Bacillariaceae, Epithemiaceae, Surirellaceae* (Bd. 2/2). Stuttgart; New York: Gustav Fischer Verlag. 596 p.

Krammer, K. & Lange-Bertalot, H. (1991a). *Süßwasserflora von Mitteleuropa: Bacillariophyceae: Centrales, Fragilariaceae, Eunotiaceae* (Bd. 2/3). Stuttgart; Jena: Gustav Fischer Verlag. 576 p.

Krammer, K. & Lange-Bertalot, H. (1991b). *Süßwasserflora von Mitteleuropa: Bacillariophyceae: Achnanthaceae, Kritische Ergänzungen zu Navicula (Lineolatae) und Gomphonema Gesamtliteraturverzeichnis* (Bd. 2/4). Stuttgart; Jena: Gustav Fischer Verlag. 437 p.

Labay, V. S., Zavarzin, D. S., Moukhametova, O. N., Konovalova, N. V., Motylkova, I. V. & Polupanov, P. V. (2010*). Plankton and benthos of Vavajskaya lakes system (Southern Sakhalin) and conditions of their dwelling*. Yuzhno-Sakhalinsk: Publishing house of SakhNIRO. 216 p.

Lange-Bertalot, H. (2001). *Diatoms of Europe. Diatoms of the European inland waters and comparable habitats: Navicula sensu stricto, 10 Genera Separated from Navicula sensu stricto, Frustulia*, (V. *2*). Ruggell: A.R.G. Gantner Verlag K.G. 526 p.

Lange-Bertalot, H. & Genkal, S. I. (1999). *Iconographia diatomologica: annotated diatom micrographs: Diatoms from Siberia I. Islands in the Arctic Ocean (Yugorsky-Shar Strait)*, (V. *6*). Königstein, Germany: Gantner Verlag K.G. 304 p.

Medvedeva, L. A. (2013). First results of algological study of Dagy River (Sakhalin Island) In E. A. Makarchenko (Ed.), *Freshwater Life* (v. *1*, 38–48). Vladivostok: Dalnauka. (In Russian).

Medvedeva, L. A. & Miski, A. V. (2011). Materials on the flora of freshwater algae from western coast of Sakhalin Island. In E. A. Makarchenko (Ed.), *Vladimir Ya. Levanidov's Biennial Memorial Meetings*, (v. *5*, 346–359). Vladivostok: Dalnauka. (In Russian).

Mikishin, Yu. A. (2008). Upper Holocene lake deposits of the bottom of the Tym River – traces of a natural disaster in northern Sakhalin. In N. G. Razshigaeva & L. A. Ganzei (Eds.), *Climatic changes, natural catastrophes and landscape development of south Far East at Pleistocene-Holocene* (86–97). Vladivostok: Dalnauka. (In Russian).

Mogilnikova, T. A., Latkovskaya, E. M. & Nikulina, T. V. (2013). Spatial variability of hydrochemical parameters and algal communities at the boundary of a river-sea. In E. A. Makarchenko (Ed.), *Freshwater Life*, (v. *1*, 212–225). Vladivostok: Dalnauka. (In Russian).

Motylkova, I. V. & Konovalova, N. V. (2003). The spring phytoplankton of Tunaicha Lake (South Sakhalin). In E. A. Makarchenko (Ed.), *Vladimir Ya. Levanidov's Biennial Memorial Meetings*, (v. *2*, 346–359). Vladivostok: Dalnauka. (In Russian).

Motylkova, I. V. & Konovalova, N. V. (2008). The summer phytoplankton of Vavay Lakes system (South Sakhalin). In E. A. Makarchenko (Ed.), *Vladimir Ya. Levanidov's Biennial Memorial Meetings*, (v. *4*, 108–117). Vladivostok: Dalnauka. (In Russian).

Motylkova, I. V. & Konovalova, N. V. (2010). Seasonal dynamics of phytoplankton in a lagoon-type lake Izmenchivoye (Southeast Sakhalin). *Russian Journal of Marine Biology*, v. *36* (2), 86–92.

Motylkova, I. V. & Konovalova, N. V. (2011). Structure of summer phytoplankton of Sladkoye Lake (Northwest Sakhalin). In E. A. Makarchenko (Ed.), *Vladimir Ya. Levanidov's Biennial Memorial Meetings* (v. *5*, 370–385). Vladivostok: Dalnauka. (In Russian).

Motylkova, I. V. & Konovalova, N. V. (2012a). Dynamic of phytoplankton from lagoon-type Tunaicha Lake (the South Sakhalin). *Hydrobiological Journal*, N5, v. *48*, 30–38. (In Russian).

Motylkova I. V. & Konovalova, N. V. (2012b). Some data of phytoplankton of the mountain lakes from plateau Spamberg (Sakhalin Island, Russia). In: S. P. Vasser (Ed.), Algologia. Supplement. Abstrsct book of IV Intrnational Conference *"Advances in modern Phycology"*, (204–205). (In Russian).

Nikulina, T. V. (2005a). Diatom algae of fresh waters from south Sakhalin. In S. I. Genkal (Ed.), Abstract book of IX International Scientific conference of Diatomologists *"Morphology, systematic, ontogeny, biogeography of diatom algae"*, (50–51). Borok; IBIW RAS. (In Russian).

Nikulina, T. V. (2005b). Diatom algae (Bacillariophyta) from the south part of Sakhalin Island. In S. Yu. Storozhenko (Ed.), *Flora and fauna of Sakhalin Island (Materials of International Sakhalin Island Project)*, (v. 2, 8–20). Vladivostok: Dalnauka. (In Russian).

Nikulina, T. V. (2009a). Diatoms of hot springs of Sakhalin Island (Far East, Russia). *Phycologia*, v. *48*, N4, 93.

Nikulina, T. V. (2009b). Structure of algal communities and water quality assessment of Tym River and Poronai River (Sakhalin Island, Russia). In A. F. Alimov & A. V. Adrianov (Eds.), *Abstract book of X Congress of Hydrobiological Society of Russian Academy of Science*, (291–292). Vladivostok: Dalnauka. (In Russian).

Nikulina. T. V. (2011a). Diatom Flora of the Tym River basin (Sakhalin Island, Russia). In O. V. Anisimova (Ed.), *Proceedings of the XII International Scientific conference of Diatomologists "Diatom algae: morphology, systematic, floristic, ecology, paleogeography, biostratigraphy"*, (118–121). Moscow: University Book. (In Russian).

Nikulina, T. V. (2011b). Spatial dynamics of periphyton algal communities and change of water quality in the Tym River basin (Sakhalin Island, Russia). In E. A. Makarchenko (Ed.), *Vladimir Ya. Levanidov's Biennial Memorial Meetings*, (v. *5*, 396–411). Vladivostok: Dalnauka.

Nikulina, T. V. & Kociolek, J. P. (2011). Diatoms from hot springs from Kuril and Sakhalin Islands (Far East, Russia) In: Seckbach J. & Kociolek J. P., editors. *The Diatom World*. London, New York: Springer. 333-363.

Pantle, F. & Buck, H. (1955). Die biologische Überwachung der Gewasser und die Darstellung der Ergebnisse. *Gas-und Wasserfach*, bd 96, N18, 1–604.

Resources of surface waters of the USSR: Far East. Sakhalin and Kurils (1973). In M. G. Vaskovskyi, (v. *18*. Is. 4). Leningrad: Gydrometizdat. 262 p. (In Russian).

Samatov, A. D., Labay, V. S., Motylkova, I. V., Mogilnikova, T. A., Zavarzin, D. S. & Ni N. K. (2002). Description of the aquatic biota of the Tunaicha Lake (South Sakhalin) in summer period. *Transactions of SakhNIRO: Water life biology, resources status and condition of inhabitation in Sakhalin–Kuril region and adjoining water areas*, v. *4*, 258–269. (in Russian).

Safronov, S. N., Litenko, N. L., Peshehodko, V. M., Labay, V. S., Stepanova, T. G. & Kolganova, T. N. (2000). Ecological and biocenotical characteristics and water quality of inland waters of Sakhalin Island. In L.V. Bushueva (Ed.), *V. V. Stanchinskyi Biennial Memorial Meetings*, (v. 3, 321-327). Smolensk: Publishing house of Smolensk University. (in Russian).

Sládeček, V. (1967). System of water quality from the biological point of view. *Archiv für Hydrobiologie Beiheft Ergebnisse der Limnologie*, v. 7, 1–218.

Sládeček, V. (1986). Diatoms as indicators of organic pollution. *Acta Hydrochimica et hydrobiologica*, v. *14*, N5, 555–566.

Sörensen, T. A. (1948). A method of establishing groups of equal amplitude in plant sociology based on similarity of species content. K. *Danske Vidensk Selk.*, *N5* (4), 1–34.

Swift, E. (1967). Cleaning diatoms frustules with ultraviolet radiation and peroxide. *Phycologia*, v. *6*, N2–3, 161–163.

Usova, N. P., Filatova, V. I. & Chernishova, E. R. (1980). On hydrobiological environment of the Tunaicha Lake. In Yu. A. Kolesnik (Ed.), *Distribution and management of water zoological resources of the Sakhalin and the Kuril Islands*, (8–16). Vladivostok: FESC AS USSR. (in Russian)

Van Dam, H., Mertens, A. & Sinkeldam, J. (1994). A coded checklist and ecological indicator values of freshwater diatoms from the Netherlands. *Netherlands Journal of Aquatic Ecology*, v. *1* (28), 117–133.

Zharkov, R. V. (2008). Daginskoe deposit of thermomineral waters in the north of the Sakhalin Island. In O.N. Likhacheva (Ed.), *Proceedings of the 2nd Sakhalin Scientific Workshop for youth. Natural accidents: studying, monitoring, forecasting*, (285–290). Yuzhno-Sakhalinsk, Sakhalin: *Institute* of Marine Geology and Geophysics. (in Russian)

In: Diatoms
Editor: Flaubert C. Bour

ISBN: 978-1-62948-210-1
© 2013 Nova Science Publishers, Inc.

Chapter III

Paleoceanogaphy since the Warm Pliocene Epoch in the Mid-Latitudes of the Northwestern Pacific Ocean

Itaru Koizumi[1]* **and Hirofumi Yamamoto[2]***

[1]Emeritus Professor of Hokkaido University, Sapporo, Hokkaido, Japan
[2]Marine Technology and Engineering Center (MARITEC),
Japanese Agency for Marine-Earth Science and Technology (JAMSTEC),
Yokosuka, Japan

Abstract

Diatom-derived sea surface temperatures (*Td′*-SST) (°C) (Koizumi, 2008) after 5.3 Ma gradually decreased with pronounced warm peaks at 4.5-4.3, 3.7-3.3, and 3.1 Ma at DSDP Sites 578-580 and Site 436 in the middle latitudes of the northwestern Pacific Ocean. At northern site 436, estimated paleotemperatures based on the comparison between *Twt* and *Td′*-SSTs (°C) have large differences from present-day mean annual temperatures ranging from 22 °C to 10°C. The values at southern Site 578, in contrast, show small fluctuations with the largest difference of 4.5°C.

* Corresponding author: Itaru Koizumi Atsubetsu-kita 3-5-18-2, Atsubetsu-ku, Sapporo 004-0073, Japan itaru@sci.hokudai.ac.jp.

Fluctuations of *Td'*-SSTs (°C) and diatom assemblages suggest influences of Earth's orbital forcing cycles.

Diatom abundances (10^7 valves/g) and *Td'*-SSTs (°C) decrease after 4.8 Ma at Site 579 in the modern Mixed Water Region. The relative abundances of subtropical-warm diatom species *Thalassionema nitzschioides* s.l. predominate occupying 40-60 % of the diatom assemblages in the early Pliocene, and 15-45 % in the late Pliocene at Site 579. And remarkable decrease occurs at 2 Ma. The relative abundances of sub-littoral upwelling diatom *Chaetoceros* spores increase by 5-20 % over after 2 Ma. During the Pliocene epoch, cold upwelling front may be absent in the subtropical-warm water region because latitudinal thermal gradient between the equator and the subtropical/ middle latitudes was very weak during the warm Pliocene epoch.

Introduction

The pre-glacial/glacial boundary occurred at 2.85-2.50 Ma (Myr ago) in the Mixed Water Region around 40° N in the western North Pacific Ocean (Figure 1).

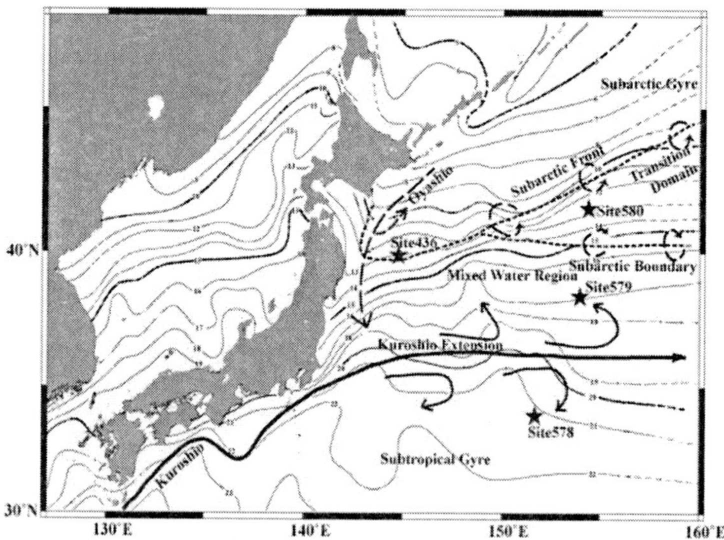

Figure 1. Map showing the location sites of four DSDP cores (stars) and schematic illustration of the pathways of the Oyashio water intrusions across the Subarctic Front into the Mixed Water Region, Transition Domain and Kuroshio Extension Region (Masujima et al., 2003) off central Japan, northwest Pacific Ocean.

Since then the surface water temperatures have never risen to the present-day level. Prior to the onset of Northern Hemisphere glaciation, significant climatic warming occurred with higher pronounced warm peaks at 4.5-4.3, 3.7-3.3, and 3.1 Ma during the early and middle Pliocene (Koizumi, 1985, 1986). Sancetta and Silvestri (1986) suggests that the present-day Subarctic Gyre did not exist prior to 2.5 Ma in the northwestern Pacific and a broad transitional area was present between 41° to 47° N in the North Pacific.

The Pliocene warm period (between 5 and 3 Ma) are widely studied as an analogue of the global warmer climate than today. Since 1990, the Pliocene Research, Interpretations, and Synoptic Mapping (PRISM) of the U.S. Geological Survey aimed to reconstruct global environmental conditions in the warm interval around 3.0 Ma prior to the onset of Northern Hemisphere glaciation (Cronin and Dowsett, 1991). In the PRISM Project, Barron (1992) analyzed diatom paleoclimatic data for the interval between 3.25 and 2.08 Ma at Site 580 using a paleoclimatic curve "*Twt*" based on the ratio of warm-water (subtropical) to cold-water diatoms with warm-water transitional taxa (*Thalassionema nitzschioides*, *Thalassiosira oestrupii*, and *Coscinodiscus radiatus*) factored into the equation at an intermediate (0.5) value. The *Twt* ratios reveal a middle Pliocene (between 3.1 and 3.0 Ma) warm interval at Site 580 during which paleotemperatures may have exceeded maximum Holocene values by 3°–5.5°C (Barron, 1992). As summarized by Dowsett et al. (1994), the middle Pliocene interval is characterized by moderate to extreme (+2 °C to more than +6°C) warming of surface waters in the middle- to high-latitude North Atlantic and North Pacific oceans.

The diatom paleotemperatures at Sites 579 and 580 respond to Earth's orbital forcing throughout the Quaternary (Koizumi, 1994). Maximum entropy spectral analyses and a fitting test to find the best suitable curve for the modified time series based on the non-linear least squares for *Td* (diatom temperature) ratio were performed for the Quaternary portion of Sites 579 and 580. Among dominant cycles during the Brunhes Normal-Polarity Chron (between 0.73 Ma and present-day). 411.5 kyr and 126.0 kyr at Site 579, and 467.0 kyr and 136.7 kyr at Site 580 correspond to 413 kyr and 95 to 124 kyr in the cycles of the orbital eccentricity. Minor cycles of 41.2 kyr at Site 579 and of 41.7 kyr at Site 580 are considered to correspond to 41 kyr in the cycles of the obliquity (tilt). During the Matuyama Reversed-Polarity Chron (between 3.18 and 0.73 Ma) at Site 580, cycles of 49.7 kyr and 43.6 kyr appear dominant.

During the interval from 4.9 to 4.1-3.7 Ma in the early Pliocene, extinct fossil freshwater diatom species *Aulacoseira praeislandica* and near-shore marine diatom species *Koizumia tatsunokuchiensis* are abundant at DSDP Site 436 on the abyssal floor far east over the Japan Trench (Koizumi and Sakamoto, 2012). The close correspondence of abundances of two characteristic diatom groups with occurrence of coarse volcanic debris, increase of eolian material, and large numbers of fecal pellets in the early Pliocene sediments, at Site 436, suggests deposition by the settled water column after being transported to the area through the Kuroshio-Kuroshio Extension system driven by winds and atmosphere by the typhoons during a warming climate interval.

The purpose of this paper is to define paleoclimatic and paleoceanographic events after 5.3 Ma based on the diatom-derived sea surface temperatures (Td'-SST) (°C) and diatom assemblages in a series of excellent biosiliceous sequences at DSDP Sites 578-580 and Site 436 recovered on a south-north transect between 34° and 42° N in the western North Pacific Ocean.

Material and Methods

We used the original database from DSDP Sites 578-580 (Koizumi and Tanimura, 1985), and Site 436 (Koizumi and Sakamoto, 2012). Site 578 is locate in the Subtropical Gyre, Site 579 in the Mixed Water Region between the Kuroshio Extension and the Oyashio Front, Site 580 in the Transition Domain Region, and Site 436 in the Subarctic Front (Figure 1).

Age (Ma) vsersus depth (m) models for the core sediments are based on the combination of paleomagnetic data (Bleil, 1985) and diatom datum levels (Motoyama and Maruyama, 1998; Koizumi and Sakamoto, 2012), as shown in Table 1 and Figure 2.

Hydrographic Condition

The ocean waters off the East coast of Japan Islands in the northwest Pacific have four different water regions (Figure 1): the Subtropical Gyre (with the Kuroshio and Kuroshio Extension flowing eastward), the Mixed Water Region off shore between 35° and 40° N, the Transition Domain with increasing flows eastward under the influences of Oyashio and Kuroshio waters, and the Subarctic Gyre (Masujima et al., 2003).

Table 1. Chronostratigraphic framework, and latitude, longitude and water depth (m) for four DSDP cores: Sub-bottom depth (m) versus sediment age (Ma) for four DSDP cores. L: last occurrence, LC: last common or consistent occurrence, F: first occurrence, T: top, B: base

Cores		Site 578	Site 579	Site 580	Site 436
Latitude (°N)		33.9	38.63	41.63	55.9
Longitude (°E)		151.6	153.83	153.98	145.6
Water depth (m)		6010	5737	5375	5240
Sub-bottom depth (m)					
Age controls	Age (Ma)				
L *Proboscia curvirostris*	0.30				37.10
B C1n (B Brunhes)	0.78	27.66	30.65	40.00	
T C1r.1n (T Jaramillo)	0.99	31.86	38.65	48.69	
B C1r.1n (B Jaramillo)	1.07	34.46	39.85	53.03	
LC *Actinocyclus oculatus*	1.24				62.60
T C2n (T Olduvai)	1.78	54.17	61.93	82.33	
B C2n (B Olduval)	1.95	58.16	65.63	89.13	
L *Neodenticula koizumii*	2.00				96.70
T C2An (T Gauss)	2.58	72.66	87.95	121.44	
LC *Neodenticula kamtschatica*	2.65				135.00
T C2An.1r (T Kaena)	3.03	80.67	98.52	139.63	
B C2An.1r (B Kaena)	3.12	82.45	103.21	143.01	
T C2An.2r (T Mammoth)	3.21	83.66	104.71	147.33	
B C2An.2r (B Mammoth)	3.33	85.06	107.92	151.06	
B C2An (B Gauss)	3.60	87.66	115.01		
F *Neodenticula koizumii*	3.74				189.20
T C3n.1n (T Cochiti)	4.19	93.41			
B C3n.1n (B Cochiti)	4.30	94.71			
T C3n.2n (T Nunivak)	4.49	96.56	136.38		
B C3n.2n (B Nunivak)	4.63	97.66	141.95		
T C3n.3n (T Sidufjall)	4.80	99.21			
B C3n.3n (B Sidufjall)	4.90	101.58			
T C3n.4n (T Thvera)	5.00	102.56			
B C3n.4n (B Thvera)	5.24	104.56			
F *Thalassiosira oestrupii*	5.49				236.80
B C3r (B Gilbert)	6.03	109.23			

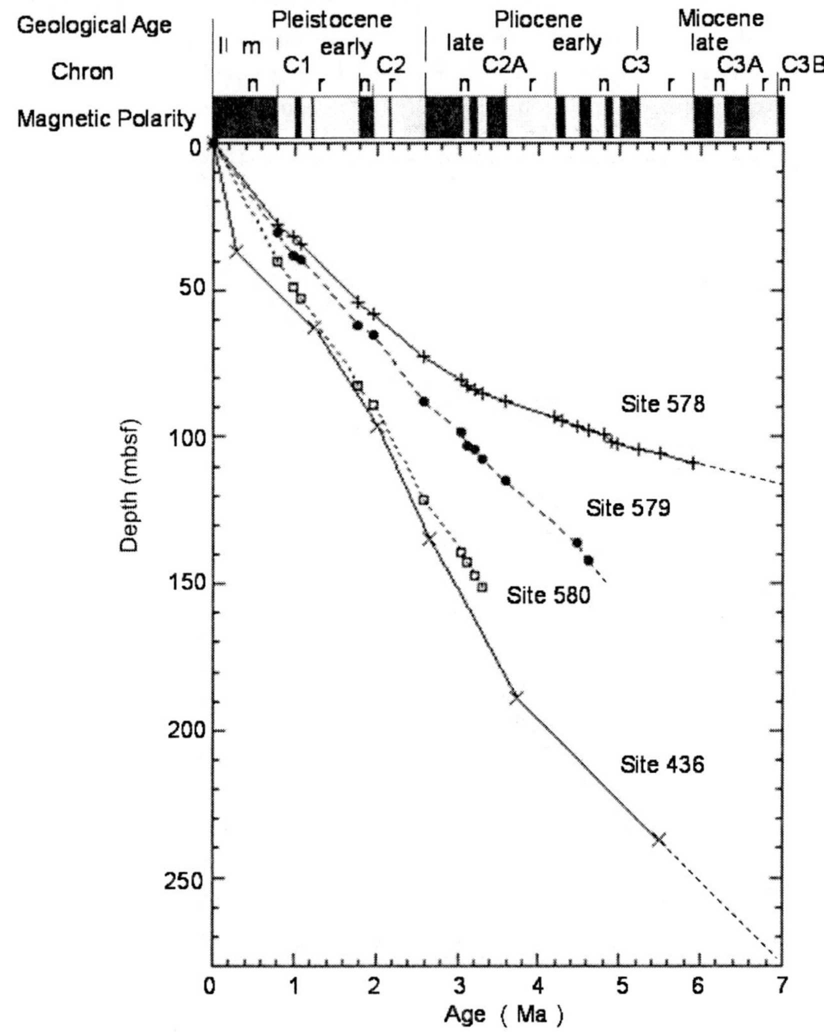

Figure 2. Sub-bottom depth (m) versus sediment age (Ma) models and sedimentation rates for four DSDP cores. Age is based on the magnetic polarity events for Sites 578-580 (Bleil, 1985) and the diatom datum levels (Motoyama and Maruyama, 1998; Koizumi and Sakamoto, 2012) after ATNTS2004 of Lourens et al. (2004) (Table 1).

The low-salinity Oyashio water flows into the Mixed Water Region and the Kuroshio Extension, forming waters with the minimum salinity, which is widely distributed and forms North Pacific Intermediate Water in the North Pacific subtropical region. The Oyashio is the western boundary current of the western subarctic gyre in the North Pacific.

The Mixed Water Region, being a part of the subtropical region, is between the northern edge of the Kuroshio Extension and the Subarctic Front. In this region numerous meanders and eddies occur especially along the northern boundary of the Kuroshio Extension. The hydrographic complexities in the regions cause local upwelling due to the isopycnal mixing between the subarctic and subtropical water masses, different velocities and vertically different salinities, and thus cause high phytoplankton productivity.

Paleotemperature Estimates

Barron (1992) proposed the *Twt* ratio to interpret the Pliocene paleoclimatic changing on the region of DSDP Site 580 in the northwest Pacific. In $Twt=(Xw+0.5Xt)/(Xc+Xt+Xw)$, Xw is the total number of subtropical to tropical (warm-water) taxa. Xt is the total number of warm transitional taxa and Xc is the total number of subarctic to arctic (cold-water) taxa. Those Xw and Xc include extinct Pliocene taxa such as *Nitzschia reinholdii, N. fossilis, N. jouseae* and *Thalassiosira convexa* as Xw, and *Neodenticula kamtschatica* and *N. koizumii* as Xc.

The *Twt* ratio are regarded as promising for Pliocene paleoclimatic studies, because *Twt* values are typically highest for the most southern Site 578, intermediate for intermediate Site 579, and lowest for northern Sites 580 and 436 (Barron, 1992; Figures 3 and 4).

The *Twt* ratio in this paper constitutes of the associate taxa which *Td'* ratio (Koizumi et al., 2004) introduced and include such extinct Pliocene taxa, in addition to those Barron (1992) adopted, as *Nitzschia miocenica, Rhizosolenia praebergonii, Thalassiosira miocenica, T. praeconvexa Xw*, and *Actinoyclus oculatus, Proboscia curvirostris, Thalassiosira nidulus* as Xc (Table 2). These taxa are considered to be corresponding to their descendent based on their morphology, paleogeographic distribution, and ecological requirements.

In order to estimate the paleo-annual SST (°C), the mean annual SST values (°C) of the observed SSTs (°C) data in summer and winter were plotted against the *Td'* ratio in 123 surface sediments around the Japanese Islands (Koizumi, 2008). The *Td'*-SSTs (°C) values substitute for the paleotemperatures (°C).

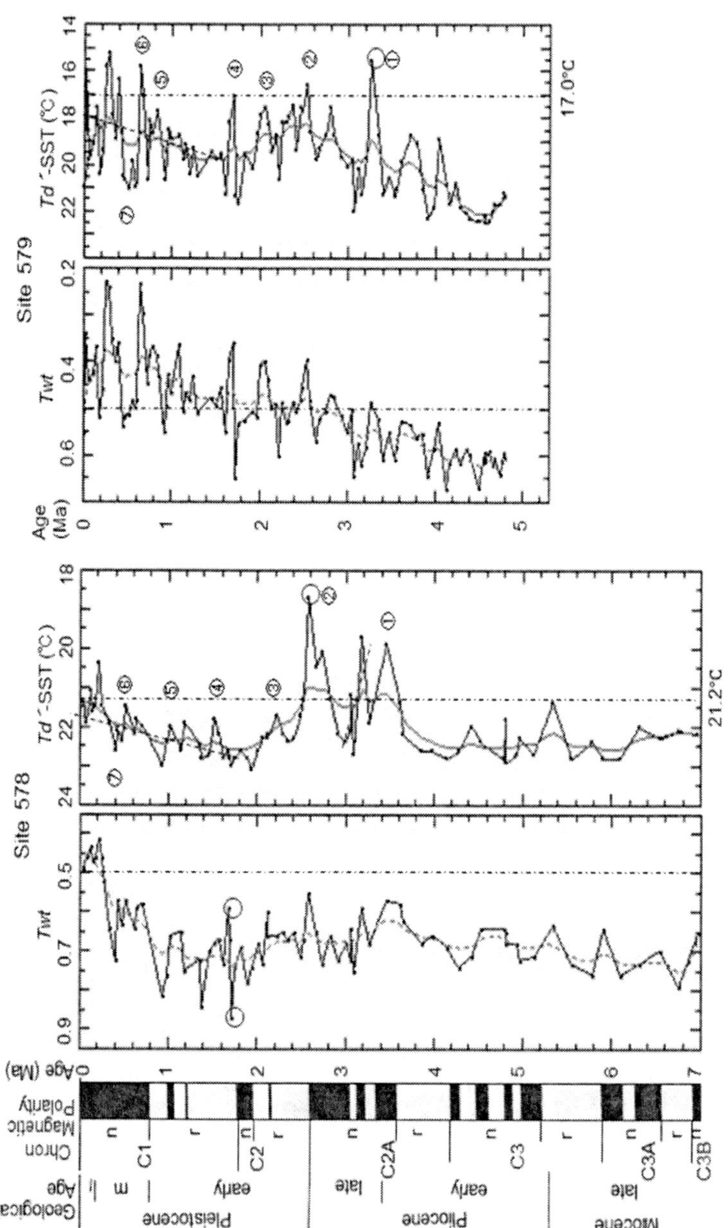

Figure 3. Comparison of sea–surface temperature records between *Twt* and *Td′*-SST (°C) at Sites 578 and 579. Circles indicate breaks away from general trends of *Td′*-SST (°C). Circled numbers represent major paleoclimatic events. Vertical broken lines represent the present-day value of sea-surface temperatures at these sites.

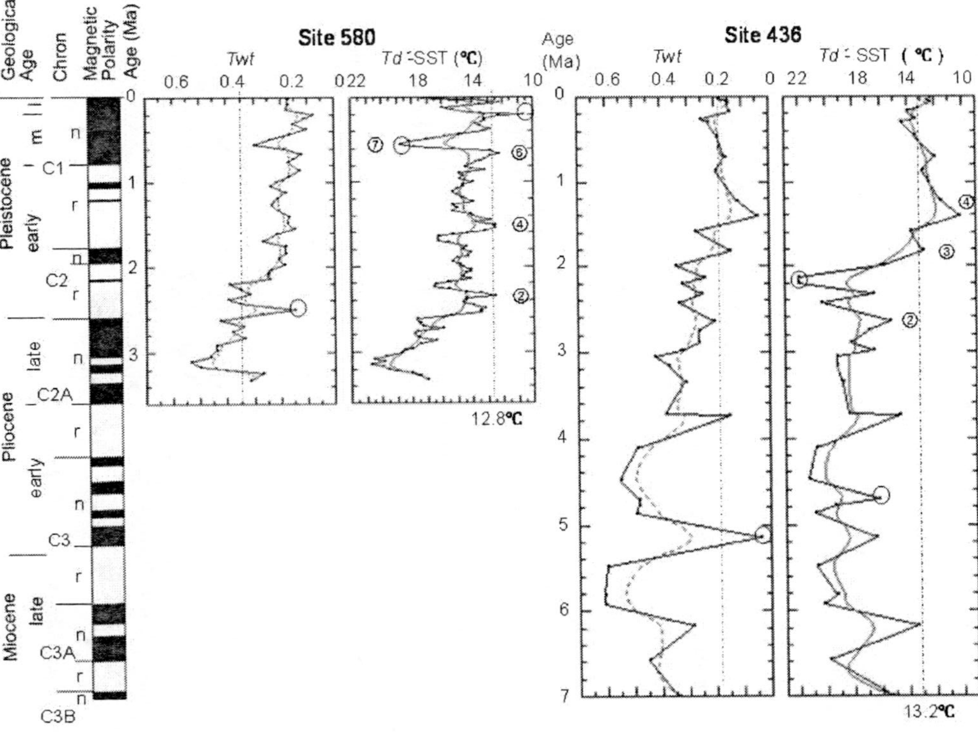

Figure 4. Comparison of sea–surface temperature records between *Twt* and *Td´*-SST (°C) at Sites 580 and 436. Circles indicate breaks away from general trends of *Td´*-SST (°C). Circled numbers represent major paleoclimatic events. Vertical broken lines represent the present-day value of sea-surface temperatures at these sites.

Table 2. Species composition for
Td' $[(Xw+XW)/(Xc+XC+Xw+XW)]$ (Koizumi et al., 2004) and Twt
$(=(Xw+0.5Xt)/(Xc+Xt+Xw))$ (Barron, 1992). Xw, XW: the total percentage
of subtropical—tropical (warm—water) taxa, Xc, XC: the total percentage
of subarctic (cold-water) taxa, Xt: warm-water transitional taxa

Td'	
Warm-water taxa (Xw+XW)	Cold-water taxa (Xc+XC)
Actinocyclus ellipticus Grunow	*Actinocyclus curvatulus* Janisch
A. elongatus Grunow *Alveus marinus* (Grunow) Kaczmarska & Fryxell	*A. ochotensis* Jousé *Asteromphalus hyalinus* Karsten
Asterolampra marylandica Ehrenberg	*A. robustus* Castracane
Asteromphalus arachne (Brebisson) Ralfs	*Bacterosira fragilis* Gran
A. flabellatus (Brebisson) Greville	*Chaetoceros furcellatus* Bailey
A. imbricatus Wallich	*Coscinodiscus marginatus* Ehrenberg
A. pettersonii (Kolbe) Thorrington-Smith	*C. oculus-iridis* Ehrenberg *Fragilariopsis cylindrus* (Grunow) Krieger
A. sarcophagus Wallich	
Azpeitia africanus (Janisch) Fryxell & Watkins	*F. oceanica* (Cleve) Hasle *N. seminae* (Simonsen & Kanaya) Akiba & Yanagisawa
A. nodulifera (Schmidt) Fryxell & Sims	
A. tabularis (Grunow) Fryxell & Sims *Fragilariopsis doliolus* (Wallich) Medlin & Sims	*Porosira glacilis* (Grunow) Jorgensen *R. hebetata* (Bailey) Gran
Hemidiscus cuneiformis Wallich	*Thalassiosira gravida* Cleve
N. interruptestriata Simonsen	*T. hyalina* (Grunow) Gran
N. kolaczekii Grunow	*T. kryophila* (Grunow) Joergensen
Planktoniella sol (Wallich) Schütt *Pseudosolenia calcar-avis* (Schültze) Sundstrom	*T. nordenskioldii* Cleve *T. trifulta* Fryxell
Rhizosolenia acuminata (Peragallo) Gran	
R. bergonii Peragallo *R. hebetata* (Bailey) Gran f. *semispina* (Hersen) Gran	
R. imbricata Brightwell	
Roperia tessellata (Roper) Grunow	
T. leptopus (Grunow) Hasle & Fryxell	
T. oestrupii (Osterfeld) Proshkina-Lavrenko	

Twt		
Warm-water taxa (*Xw*+XW) of *Td'* + fossil warm-water taxa	Warm-water transitional taxa (*Xt*)	Cold-water taxa (*Xc*+XC) of *Td'* + fossil cold-water taxa
Nitzschia fossilis (Frenguelli) Kanaya *N. jouseae* Burckle *N. miocenica* Burckle *N. reinholdii* Kanaya *R. praebergonii* Mukhina *Thalassiosira convexa* Mukhina *T. miocenica* Schrader *T. praecovexa* Burckle	*Coscinodiscus radiatus* Ehrenberg *Thalassionema nitzschioides* Grunow s.l. (convert from *Xw* of *Td'*) *Thalassiosira oestrupii* (Osterfeld) Proshkina-Lavrenko	*Actinocyclus oculatus* Jousé *Neodenticula kamtschatica* (Zabelina) Akiba & Yanagisawa *N. koizumii* Akiba & Yanagisawa *Proboscia barboi (*Brun) Jordan & Priddle *P. curvirostris* (Jousé) Jordan & Priddle *Thalassiosira nidulus* (Tempère & Brun) Jousé

Diatom Assemblages and Hydrographic Conditions

At Site 579 where local upwelling causes high phytoplankton productivity, the diatom abundances and diatom taxa were quantitatively analyzed. The dominant taxa throughout the record at Site 579 are *Thalassionema nitzschioides* sensu lato (s.l.) and *Chaetoceros* spores. The occurrences of *Thassionema-Thalassiothrix* ooze in the frontal boundary of the eastern equatorial Pacific (Kemp and Baldauf, 1993; Kemp et al., 1995) and subarctic North Pacific (Dickens and Barron, 1997) are restricted in the latest Miocene–early Pliocene.

A close relationship between high concentrations of *Chaetoceros* spores, high productivity, and intensive upwelling recognize in the surface sediments of the Bering and Okhotsk seas (Sancetta, 1982). *Chaetoceros* spores are a characteristic component of the flora during late-stage upwelling with rapid aggregation and sinking in many coastal regions (Sancetta et al., 1992; Kemp et al., 2000).

Results and Discussion

Sedimentation Rates

The sedimentation rate at Site 436, far eastern and near-shore over the Japan Trench, increase from 30 m/Myr in the late Miocene to early Pliocene (7.0 to 3.7 Ma) to 55 m/Myr at 190 m (3.5 Ma, early/late Pliocene boundary), approximately two times higher, and continued to be high through the late Pliocene and early Pleistocene (3.7 to 1.3 Ma) (Koizumi and Sakamoto, 2012). At northern Site 580, the sedimentation rate is also high in the Pliocene and Pleistocene, averaging 50 m/Myr. On the other hand, at Site 578, southern and off-shore in the Subtropical Gyre, the sedimentation rate is 10 m/Myr in the late Miocene to Pliocene (7.0 to 2.6 Ma). At 76 m (2.6 Ma), it increase to 30 m/Myr, and continue to be in higher rates through the Pliocene and Pleistocene after 2.6 Ma. The upper section 76 m is anoxic gray to olive gray colored biosiliceous clays with many pyrite-cemented layers. The areas of high diatom productivity shifted from the northeast to the northwest Pacific Ocean, and productivity became higher by enhanced upwelling along the coastal regions due to rapid cooling at 2.7 Ma (Barron, 1998).

Comparison of SSTs

The *Twt* curves are compared with *Td'*-SSTs (^{o}C) curves for four DSDP sites (Figures 3 and 4). Differences between two kinds of curves are not remarkable, but *Twt* ratio shows consequences from the relative abundances of extinct Pliocene warm- and/or cold-water species, while *Td'*-SSTs (^{o}C) curve does not.

At Site 578, *Td'*-SSTs (^{o}C) show the lower temperature 18.7 °C, because warm-water Pliocene species *Nitzschia fossilis* (relative abundance 13.5 %) does not occur at 2.57 Ma (Figure 3). The predominant occurrences of *N. reinholdii* (59.6 %) pull up *Twt* ratio at 1.71 Ma, while the decreases of *N. fossilis* and *N. reinholdii* pull down at 1.67 Ma. The values of *Td'*-SSTs (^{o}C) fluctuate relatively flat, compared with *Twt* ratio. At Site 579, the *Td'*-SSTs (^{o}C) drop down to 15.6 °C at 3.26 Ma indicating due to the exclusion of *N. reinholdii* (16.0 %).

At Site 580, *Twt* ratio reduces due to the predominance of *Neodenticula koizumii* (26.3 %) at 2.50 Ma. The *Td'*-SSTs (^{o}C) show higher temperature 18.8^{o}C due to predominance of warm-water species *Thalassiosira oestrupii*

(53.0 %) at 0.55 Ma, while the increases both of cold-water species *Neodenticula seminae* (42.7%) and *Thalassiosira gravida* (11.6 %) pull down at 0.21 Ma. At Site 436, *Twt* ratio remarkably drop down due to the predominance of *Coscinodiscus marginatus* (59.7 %) at 5.15 Ma. The *Td'*-SSTs (ºC) drop at 4.71 Ma due to decrease of *Azpeitia nodulifera* in spite of exclusion of *Neodenticula kamtschatica*, while the predominances both of cold-water extinct species *N. koizumii* (36.1-61.5 %) and *Rhizoslenia barboi* pull up temperatures at 2.21-2.14 Ma.

Major Paleoclimatic Events

At four DSDP sites, the *Td'*-SSTs (ºC) fluctuates in a similar way and generally decrease during the period from the bottom late Miocene-early Pliocene to the top of present-day. The remarkable variation of *Td'*-SSTs (ºC), common on at four sites are approximately at 23-kyr, indicating the correspondence to the precession cycle. The *Td'*-SSTs (ºC) during the last 5 Myr show five conspicuous episodes, which indicate surface water cooling: (1) Sharp drop occurs at 3.20 Ma (Site 578) and 3.26 Ma (Site 579). Initiation of Northern Hemisphere glaciation is suggested at this time (Kleiven et al., 2002; Bartoli et al., 2005). (2) A large decrease occurs at 2.57 Ma (Site 578), 2.53 Ma (Site 579), 2.35 Ma (Site 580), and 2.65 Ma (Site 436). These time levels approximately coincide with the lithologic change from oxidized clay to anoxic biosiliceous clay at each site, and indicate the beginning of the glacial age defined as the Pliocene and Pleistocene boundary. (3) The dissolution of diatom valves is observed around 2.23 Ma at Site 578. *Td'*-SSTs (ºC) decrease at 2.05 Ma at Site 579. The drop of *Td'*-SSTs (ºC) at Site 436 occurs from 2.14 Ma to 1.84 Ma, and *Td'*-SSTs (ºC) remain lower than the present-day value since then. This cooling event corresponds with the second step by Ravelo et al. (2004). (4) *Td'*-SSTs (ºC) decrease at 1.70 Ma (Site 579), 1.55 Ma (Site 580), and 1.41 Ma (Site 436). (5) Large and rhythmic fluctuations of *Td'*-SSTs (ºC) begin at 1.18 Ma (Site 578), 0.96 Ma (Site 579), and 1.24 Ma (Site 580). (6) Low-frequency and high-amplitude in fluctuations occur at 0.64 Ma (Site 578), 0.84 Ma (Site 579), and 0.85 Ma (Site 580). (7) Remarkably high *Td'*-SSTs (ºC) occur at 0.40 Ma (Site 578), 0.51 Ma (Site 579), and 0.55 Ma (Site 580). Global climatic change around 0.4 Ma, Mid-Brunhes Event (Jansen et al., 1986), was forced by the orbital eccentricity cycle. After this event, climatic trend change towards more glacial conditions in the Northern Hemisphere.

The wavelet analyses for *Td'*-SSTs (°C) at Site 579 in the Mixed Water Region indicates pronounced variability with duration 123 to 95-kyrs during 4.8 Ma, suggesting the correspondence to the eccentricity cycle. Variability of 54 to 41-kyr periods suggests the correspondence to the orbital obliquity cycles in five intervals; 3.3-3.0 and 1.8-1.4 Ma, 800-600, 400, and 100 ka (Figure 6). During 4.1-2.6 Ma and 800-200 ka, variability of 410-kyr period is recognized, also variability of 250-kyr period during the intervals of 4.1-3.8 and 3.5-2.2 Ma.

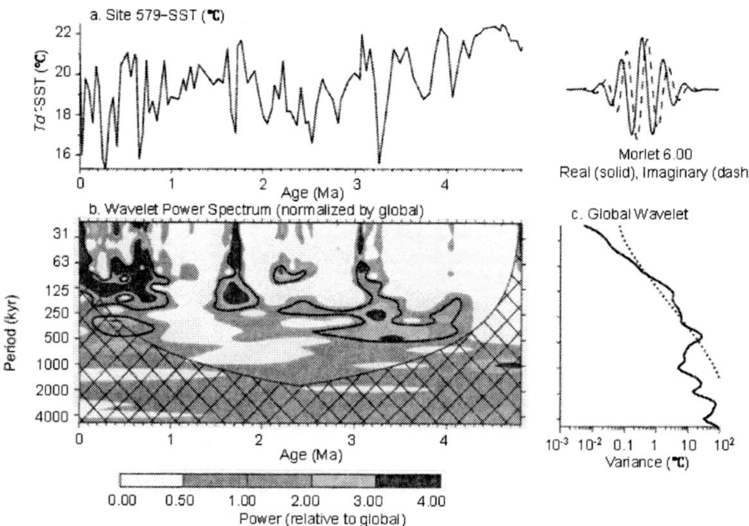

Figure 5. The wavelet analysis (Torrence and Compo, 1998) for the values of the band ratio annual *Td'*-SST (°C) since 4.8 Ma at Site 579. a: Annual *Td'*-SST (°C) since 4.8 Ma at Site 579. b: Wavelet power spectrum. Solid black line indicates the 10 % confidence level (power significance). The contour levels are chosen for 75 %, 50 %, 25 %, and 5 % of the wavelet power, respectively. More intense colors indicate higher power. c: The global wavelet power spectrum (black line). The dashed line indicates the significance for the global wavelet spectrum, assuming the same significance level and background spectrum as in b. The cross-hatched region is the cone of influence, where zero padding has reduced the variance.

Diatom Abundances and Assemblages

Higher diatom concentrations than average value (3.7×10^7 valves/g) are recognized within the lower Pliocene clayey diatom ooze to upper Pliocene and lower Pleistocene clayey siliceous ooze. Diatom abundances decrease fluctuating with shorter durations from the bottom to the top of sequences. This decrease corresponds with an overall decline in the sea surface water temperatures from 22 °C at 4.6 Ma to 17 °C at the present-day, as Td'-SSTs (°C) indicates (Figure 6).

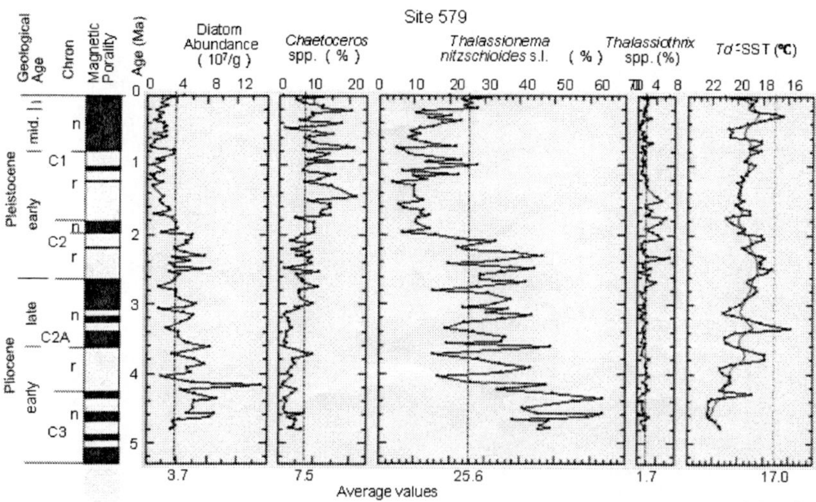

Figure 6. Stratigraphic variation of diatom abundance (10^7 valves/g), relative abundances (%) of *Chaetoceros* spp., *Thalassionema nizschioides* s.l., *Thalassiothrix* spp., and Td'-SST (°C) at Site 579. Vertical dashed lines represent the average values of Td'-SST (°C).

A spike-like increase of diatom abundance (15.1×10^7 valves/g) at 4.17 Ma was resulted from the sudden increase of *Thalassionema nitzschioides* s.l. (48 %). In the wavelet analyses pronounced variabilities are recognized: variability of 410-kyr period suggests the correspondence to the eccentricity cycle, and variability of 54 to 41-kyr period suggests the correspondence to the orbital obliquity cycles (Figure 7). In the intervals of 4.7-4.5, 3.2-3.0, and 2.6-1.8 Ma, variability of 125 to 105-kyr period is prominent, and suggest correspondence to the eccentricity cycle. And the orbital obliquity cycles are detected in the intervals: 4.6-4.5, 4.4-4.1, 3.15-2.95, 2.55-2.20, 2.05-1.95, 1.7, and 1.0-0.9 Ma.

The sub-littoral upwelling diatom *Chaetoceros* spores occur in small quantities under average value (7.5 %) throughout the early to late Pliocene section below 3.0 Ma, and then slightly increase in the upper section toward 2.0 Ma. The numbers of *Chaetoceros* spores fluctuate wide occupying from 7.5 % to 20 % in the upper section toward 2.0 Ma. After 1.7 Ma, the pronounced variability of 250 to 125-kyr period is detected by the wavelet analyses.

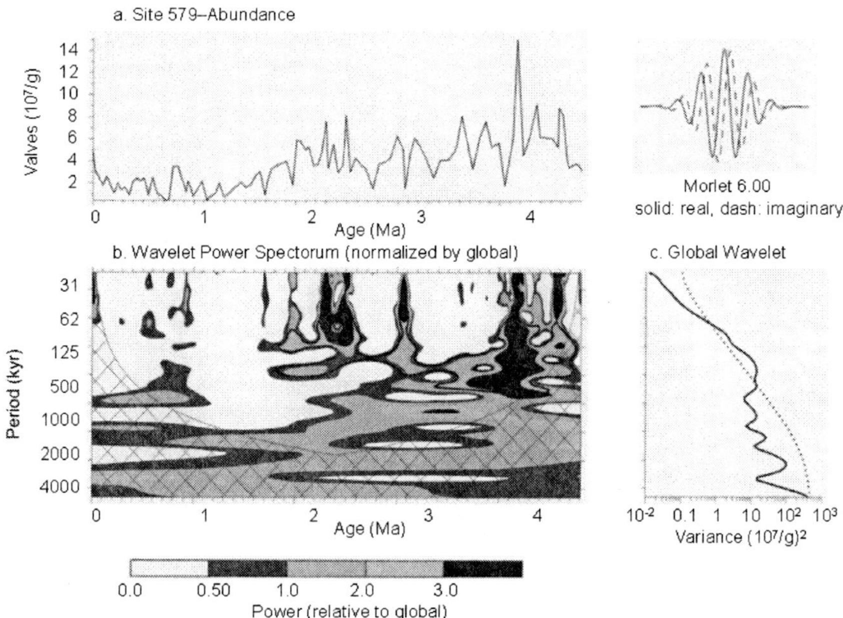

Figure 7. The wavelet analysis (Torrence and Compo, 1998) for the values of the band ratio diatom abundance (10^7 valves/g) since 4.8 Ma at Site 579. a: Diatom abundance (10^7 valves/g) since 4.8 Ma at Site 579. b: Wavelet power spectrum. Solid black line indicates the 10 % confidence level (power significance). The contour levels are chosen for 75 %, 50 %, 25 %, and 5 % of the wavelet power, respectively. More intense darkness indicate higher power. c: The global wavelet power spectrum (black line). The dashed line indicates the significance for the global wavelet spectrum, assuming the same significance level and background spectrum as in b. The cross-hatched region is the cone of influence, where zero padding has reduced the variance.

At Site 579, the relative abundances of subtropical-warm diatom *T. nitzschioides* s.l. predominate over the average value (25.6 %) throughout the section of the Pliocene and lower Pleistocene to 2.0 Ma. The valves decrease during the interval between 2.0 and 1.2 Ma, but turn to increase and occupy 30 % following the rise in temperature in the upper section to 1.5 Ma (Figure 6).

Thelassiothrix spp., which is a key "fall dump" genus and also involved in massive sedimentation beneath oceanic frontal zones (Kemp et al., 2000), occurs in a small quantity around 2.0 Ma (between 2.4 and 1.3 Ma), making a minor element, at Site 579.

Cooling steps are recognized twice since 4 Ma in the subtropical/mid-latitudes (Ravelo et al., 2004). The first step (between 3.0 and 2.5 Ma) was coincident with the onset of significant North Hemisphere glaciation. The second step (between 2.0 and 1.5 Ma) caused the modern mode of circulation with relatively strong Walker circulation and cool subtropical/mid-latitudes temperatures. The increase in the latitudinal temperature gradient at the first cooling step caused the increase in wind-eccentricity, driven subtropical upwelling, and then the western tropical warm pool was enhanced just after the second cooling step.

Conclusion

The sea surface temperatures (SSTs) (^{0}C) remarkably fluctuate over 5.3 Myr both in the northern and southern sites in the mid-latitudes of the northwestern Pacific Ocean. At Site 578 in the Subtropical Region, the difference of SSTs (^{0}C) between values in 4.6 Ma and the present is 4.5°C while at Site 436 near the Subarctic Front, the difference reaches to 12°C. At Site 579 in the Mixed Water Region, the SSTs (^{0}C) are higher than the present by 5.5 to 5.0°C after 4.6 Ma, as the Td'-SSTs (^{0}C) indicates, correspondingly to high SSTs (^{0}C), 22.5°C at 4.6 Ma, 23.3^{0}C at 3.8 Ma, and 22.0°C at 3.0Ma. On the other hand, at northern Site 436, Td'-SSTs (^{0}C) increase, with more higher difference, 8.1-6.0 °C: Td'-SSTs show 20.9°C at 4.9 Ma, 21.3°C at 4.5 Ma, and 19.2°C at 3.0 Ma.

The variability of Td'-SSTs (^{0}C) and the composition of diatom assemblages are considered to have been affected by Earth's orbital forcing: eccentricity, obliquity (tilt), and precession. Around 2.0 Ma, when the latitudinal thermal gradient between the equator and subtropical/mid-latitudes made strong, and also just after the thermal gradient crossing west–to–east along the equator was formed, major component of diatom assemblages changed from subtropical-warm neritic diatom *Thalassionema nitzschioides* s.l. to sub-littoral upwelling diatom *Chaetoceros* spores, which is related to the primary productivity.

Acknowledgment

We gratefully acknowledge Drs. Naomi Harada of the Japan Agency for Marine-Earth Science and Technology (JAMSTEC) and Tsuyoshi Watanabe of Hokkaido University for critical reviews and suggestions for improving the manuscript.

References

Barron, J.A. (1992). Pliocene paleoclimatic interpretation of DSDP Site 580 (NW Pacifc) using diatoms. *Marine Micropaleontology*, 29, 23-44.

Barron, J.A. (1998). Late Neogene changes in diatom sedimentation in the North Pacific. *Journal of Asian Earth Sciences*, 16, 85-95.

Bartoli, G., Sarnthein, M., Weinelt, M., Erlenkeuser, H., Garbe-Schönberg, D., & Lea, D.W. (2005). Final closure of Panama and the onset of northern hemisphere glaciation. *Earth and Planetary Science Letters*, 237, 33-44.

Bleil, U. The magnetostratigraphy of northwest Pacific sediments, Deep Sea Drilling Project Leg 86. In: Heath, G.R., Burckle, L.H. & others editors. *Initial Reports of the Deep Sea Drilling Project*, 86. Washington, DC: U.S. Government Printing Office; 1985; 441-458.

Cronin, T.M. & Dowsett, H.J. (1991). Pliocene Climates. *Quaternary Science Review*, 10, 1-296.

Dickens, G.R. & Barron, J.A. (1997). A rapidly deposited pennate diatom ooze in Upper Miocene-Lower Pliocene sediment beneath the North Pacific polar front. *Marine Micropaleontolgy*, 31, 177-182.

Dowsett, H., Thompson, R., Barron, J., Cronin, T., Flemiong, F., Ishman, S., Poore, R., Willard, D. & Holtz, T., Jr. (1994). Joint investigations of the Middle Pliocene climate I: PRISM paleoenvironmental reconstructions. *Global and Planetary Change*, 9, 169-195.

Jansen, J.H.F., Kuijpers, A. & Troelstra, S.R. (1986). A mid-Brunhes climatic event: Long-term changes in global atmosphere and ocean circulation. *Science,* 232, 619-622.

Kemp, A.E.S. & Baldauf, J.G. (1993). Vast Neogene laminated diatom mat deposits from the eastern equatorial Pacific Ocean. *Nature,* 362, 141-144.

Kemp, A.E.S., Baldauf, J.G. & Pearce, R.B. (1995). Origins and paleoceanographic significance of laminated diatom ooze from the eastern equatorial Pacific Ocean (Leg 138). In: Pisias, N.G., Mayer, L.A.,

Janecek, T.R., Palmer-Julson, A. & van Andel, T.H. editors. *Proceedings of the Ocean Drilling Program, Scientific Results,* 138. College Station, TX: Ocean Drilling Program, 647-663.

Kemp, A.E.S., Pike, J., Pearce, R.B. & Lange, C.B. (2000). The "Fall dump"– a new perspective on the role of a "shade flora" in the annual cycle of diatom production and export flux. *Deep-Sea Research II,* 47, 2129-2154.

Kleiven, H.F., Jansen, E., Fronval, T. & Smith, T.M. (2002). Intensification of Northern Hemisphere glaciations in the circum Atlantic region (3.5-2.4 Ma) – ice-rafted detritus evidence. *Palaeogeographgy, Palaeoclimatology, Palaeoecology,* 184, 213-223.

Koizumi, I. & Sakamoto, T. (2012). Allochtonous diatoms in DSDP Site 436 on the abyssal floor off northeast Japan. *JAMSTEC Report of Research and Development,* 14, 27-38.

Koizumi, I. & Tanimura, Y. (1985). Neogene diatom biostratigraphy of the middle latitude western North Pacific, Deep Sea Drilling Project Leg 86. In: Heath, G.R., Burckle, L.H. & others editors. *Initial Reports of the Deep Sea Drilling Project,* 86. Washngton, DC: U.S. Goverment Printing Office, 269-300.

Koizumi, I. (1986). Late Neogene diatom temperature record in the northwest Pacific Ocean. *Science Reports of College of General Education,* Osaka University, 34, 145-153.

Koizumi, I. (1994). Spectral analysis of the diatom paleotemperature records at DSDP Sites 579 and 580 near the subarctic front in the western North Pacific. *Palaeogeography, Palaeoclimatology, Palaeoecology,* 108, 475-485.

Koizumi, I. (2008). Diatom-derived SSTs (Td' ratio) indicate warm seas off Japan during the middle Holocene (8.2-3.3 kyr BP). *Marine Micropaleontology,* 69, 263-281.

Koizumi, I. Late Neogene paleoceanography in the western North Pacific. In: Heath, G.R., Burckle, L.H. & others editors. *Initial Reports of the Deep Sea Drilling Project,* 86. Wshington, DC: U.S. Government Printing Office; 1985; 429-438.

Koizumi, I., Irino, T. & Oba, T. (2004). Paleoceanography during the last 150 kyr off central Japan based on diatom floras. *Marine Micropaleontology,* 53, 293-365.

Lourens, L., Hilgen, F., Shackleton, N.J., Lasker, J. & Wilson, J. Appendix 2: Orbital tuning calibrations and conversions for the Neogene Period. In: Gradstein, F., Ogg, J. & Smith, A. editors. *A Geologic Time Scale 2004.* Cambridge: Cambridge University Press; 2004; 469-484.

Masujima, M., Yasuda, I. , Hiroe, Y. & Watanabe, T. (2003). Transport of Oyashio Water across the Subarctic Front into the Mixed Water Region and formation of NPIW. *Journal of Oceanography*, 59, 855-869.

Motoyama, I. & Maruyama, T. (1998). Neogene diatom and radiolarian biochronology for the middle-to-high latitudes of the Northwest Pacific region: Calibration to the Cande and Kent's geomagnetic polarity time scales (CK 92 and CK 95). *Jour. Geol. Soc.* Japan, 104, 171-183 (in Japanese with English abstract).

Ravelo, A.C., Andreasen, D.H. Lyle, M. Lyle, A.O. & Wara, M.W. (2004). Regional climate shifts caused by gradual global cooling in the Pliocene epoch. *Nature,* 429, 263-267.

Sancetta, C. & Silvestri, S.M. (1986). Pliocene-Pleistocene evolution of the North Pacific ocean-atmosphere system, interpreted from fossil diatoms. *Paleocenography*, 1, 163-180.

Sancetta, C. (1982). Distribution of diatom species in surface sediments of the Bering and Okhotsk seas. *Micropaleontology*, 28, 221-257.

Sancetta, C., Lyle, M., Heusser, L., Zahn, R. & Bradbury, J.P. (1992). Late-glacial to Holocene changes in winds, upwelling, and seasonal production of the northern California current system. *Quaternary Research*, 38, 359-370.

Torrence, C. & Compo, G.P. (1998). A practical guide to wavelet analysis. *Bull. Amer. Meteorol. Soc.,* 79, 61-78.

In: Diatoms
Editor: Flaubert C. Bour

ISBN: 978-1-62948-210-1
© 2013 Nova Science Publishers, Inc.

Chapter IV

Microphytobenthic Community Dynamics in Intertidal Flats of Nakdong River Estuary, South Korea

***Guo-Ying Du**[1,*] **and Ik Kyo Chung**[2]*
[1]Bioengineering Department, College of Marine Life Science,
Ocean University of China, China
[2]Division of Earth Environmental System,
Pusan National University, Korea

Abstract

The biomass dynamics of microphytobenthos (MPB) in the intertidal flats of the Nakdong River estuary (South Korea) were studied from August 2006 to August 2008. On spatial and temporal distribution, the MPB exhibited evident seasonal variation among different sampling sites. Muddier sediments possessed higher biomass which had several peaks in spring, summer or winter, while in the sandiest sediment, MPB biomass was lowest and had only one peak in summer. Multivariate correlation analysis revealed that the biomass was positively correlated with mud ($<$ 63 μm) and very fine sand (63–125 μm) composition, and negatively

---

[*] Corresponding author: Gu-Ying Du. E-mail: dgydou@gmail.com.

related to fine sand (125–250 μm) and medium sand ($\geq$ 250 μm) composition, but not statistically related to dissolved inorganic nutrient concentration (NH_4^+, $NO_2^- + NO_3^-$, PO_4^{3-}, and $Si(OH)_4$), salinity and light intensity. On vertical distribution, MPB biomass declined exponentially with depth, and also related to sediment type. In muddier sediments, the exponential curve slope was larger than sandier sediments, implying lower resuspending and burying speed and higher stability of sediments. Diel vertical migrations presented a certain degree of rhythm following tidal and light cycles, viz. surfacing in low-tide emersion at daytime, migrating down at nighttime and/or high-tide submersion. However, MPB migratory rhythm also primarily determined by their biomass, which dramatically weakened the rhythm when visibly thick biofilms formed in winter: MPB cells remained at the surface, even at night and/or during high tide. Generally, our study revealed MPB community dynamics in intertidal flats as results from interactions between biotic and environmental factors.

Introduction

In estuarine intertidal ecosystems, Microphytobenthos (MPB) are important primary producers and may contribute up to 50% of total primary production (Underwood and Kromkamp, 1999; Serôdio and Catarino, 2000; Perissinotto et al., 2002). The dynamics of microphytobenthic biomass are often related to a complex set of interactions between MPB and numerous abiotic and biotic factors (MacIntyre et al., 1996; Underwood et al., 1998; Underwood and Kromkamp, 1999).

With respect to the temporal variation of MPB community, the seasonal and interannual variability mainly relate to the factors that affect the growth rate and /or health of the MPB, such as light, temperature, nutrients, salinity, and water contents, and also relate to the removal processes, such as grazing, resuspension by physical factors (tidal currents, wind-induced circulation and wave action) (Underwood, 1994; Thornthon et al., 2002; Kromkamp et al., 2006; Méléder et al., 2007).

With respect to the spatial distribution of MPB, it must be considered in three dimensions, the planar and the vertical distribution. Factors found to structure planar distribution on small- or large-scale include geomorphological features (e.g. sand ripples), salinity, nutrients, tidal exposure time and sediment type (Brotas et al., 1995; Guarini, et al., 1998; Blanchard, et al., 2000; Christie, et al., 2000; Du, et al., 2009). Factors affecting the vertical distribution could be defined as passive and active factors.

The passive factors include re-suspension, mixing and burial processes determined by hydrodynamic forcing and bedform morphology, and their effect is more apparently at large scales (Brotas and Serôdio, 1995; Lucas et al. 2000; de Brouwer et al., 2000). The active factor is the migration of MPB, which is induced by light and nutrients for their physiological requirement, and cause the short-term variability of MPB biomass in surface sediment (within 0~2 or 5mm) (Round and Palmer 1966; Kingston, 2002; Sauer et al., 2002).

This study aimed to investigate the spatio-temporal distribution of microphytobenthic community, and figure out the relationship of community dynamics with the environmental factors. And through the monthly investigation on MPB biomass variation, to model chlorophyll *a* vertical distribution, and analyze the main affecting factors; through the hourly investigation over diurnal cycle, to ascertain the effect of migration on vertical distribution.

Materials and Methods

Study Site

The Nakdong River estuary (Lat. 35°03'~04'N; Long.128°51'~57'E) is the second largest river estuary in Korea. It is a micro-tidal estuary with semidiurnal tides, having less than 2 m tidal amplitudes and about 40 km^2 intertidal flats. The estuary has a well-developed delta with sand barriers. These sand barriers, as well as the intertidal flats, continuously expand due to sediment deposition which are regulated by the waves from the open sea and the discharge of the Nakdong River.

Four sampling sites were chosen on the bare intertidal flats of the main sand barriers (sites A, B, C, and D; Figure 1). Sites A and B are located on the inside of the intertidal flats with a gentle bathymetric gradient and lower hydrodynamics, and near to oyster and laver aquaculture farms, respectively. Sites C and D are located on verge of the flats, adjoin to *Scirpus triqueter* beds, which grows from July to September. Comparing with other sites, site D is facing heavier hydrodynamics due to its location on the east shore of the main river flows into the open sea.

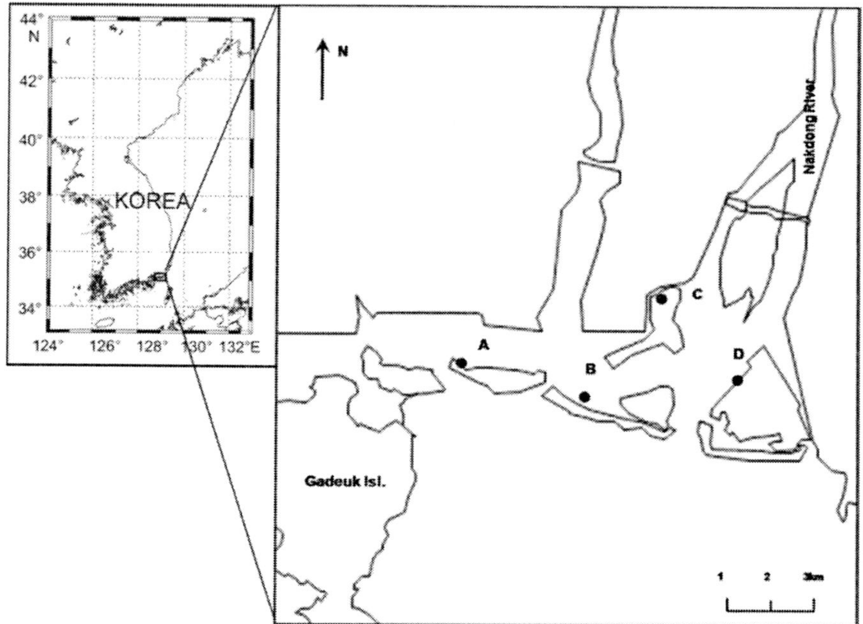

Figure 1. Location of the four study sites (A-D) in the Nakdong River estuary.

Sampling and Processing

Monthly sampling was carried out from August 2006 to August 2008, during ca. 1.5~2.0 hours before the end of the ebb tide in the afternoon at the four sites (A-D) on the same day.

Three sediment cores were collected by hand with a plexiglass corer 8 cm in diameter and 20 cm long, and sectioned into 1 cm slices down to 10 cm depth (after August 2007, only upper 1 cm was sectioned). Subsamples were taken using 50 ml polyethylene centrifuge tubes (2.7 cm in diameter), after which 15 ml of 90% acetone was added to the tube to extract pigments for measurement of chlorophyll *a* concentration. In laboratory, Chlorophyll *a* concentration was determined spectrophotometrically correcting for phaeopigments (Lorenzen, 1967).

Diel sampling was carried out at site A on 7th–28th April 2007, and 19th–20th January 2008, at ca. 1.5 hour intervals during the entire tidal period. Samples were taken using three sawn-off 50 ml tubes, and frozen by Liquid N_2 down to about 2 cm depth of sediment.

Each 0.5 mm slice of the top 1 cm was sectioned using a freezing microtome (Leitz) and extracted by 5 ml of 90% acetone for chlorophyll *a* measurement.

Light intensity and the sediment temperature (2-3 cm depth) were measured when sampling at every site, also referred the data supplied by the Korea Meteorological Administration. The salinity and dissolved inorganic nutrient concentration (NH_4^+, $NO_2^- + NO_3^-$, PO_4^{3-}, and $Si(OH)_4$) were determined in laboratory after filtering, according to the methods of Parsons et al. (1984) and the Joint Global Ocean Flux Study Protocols (1994). Sediment grain size was analyzed by four categories (medium sand ($\geq$ 250 μm), fine sand (125–250 μm), very fine sand (63–125 μm), and mud (< 63 μm)), using the dry-sieving technique (Eleftheriou and McIntyre, 2005; detailed in Du et al. 2009).

Data Analysis

The correlations between biomass and environmental factors were determined by multivariate correlation analyses. These analyses were performed using SPSS 12.0. Regression curves of the vertical distribution of chlorophyll a in the sediment were conducted using Sigma plot 8.0, following the equation of $C_Z = C_0 \times \exp(-b \times z)$ (modified from Brotas and Serôdio, 1995; $C_Z = C_0 \times \exp(-(k/v) \times z)$),where the C_Z and C_0 = chlorophyll *a* concentration at depth z and 0 ($\mu g\ cm^{-3}$ or $\mu g\ g^{-1}$); b is the reciprocal of depth (cm^{-1} or mm^{-1}), a parameter which related to the specific degradation rate of chlorophyll *a* and mean burial velocity. Contours of MPB diel vertical distribution were also made by Sigma plot.

Results and Discussion

Seasonal Variation of Biomass

From August 2006 to August 2008, distinct variation patterns of MPB biomass exhibited in different types of sediment. At sites A and B, the biomass depicted a trend where biomass was higher in winter with noticeable biofilm and lower in summer(Figure 2, a, b). Conversely, at site D in the sandy sediment, biomass was higher in summer and lower in other seasons (Figure 2,d).

It was indicated by statistical analyses that the biomass was negatively correlated to temperature at sites A and B, while positively at site D (Table 1).

MPB Biomass varied greatly at site C without any obvious seasonal trends. Several higher values at site C were observed in all four seasons with visible MPB biofilms presented (Figure 2, c).

Also there was no significant relationship of biomass at site C with temperature (Table 1).

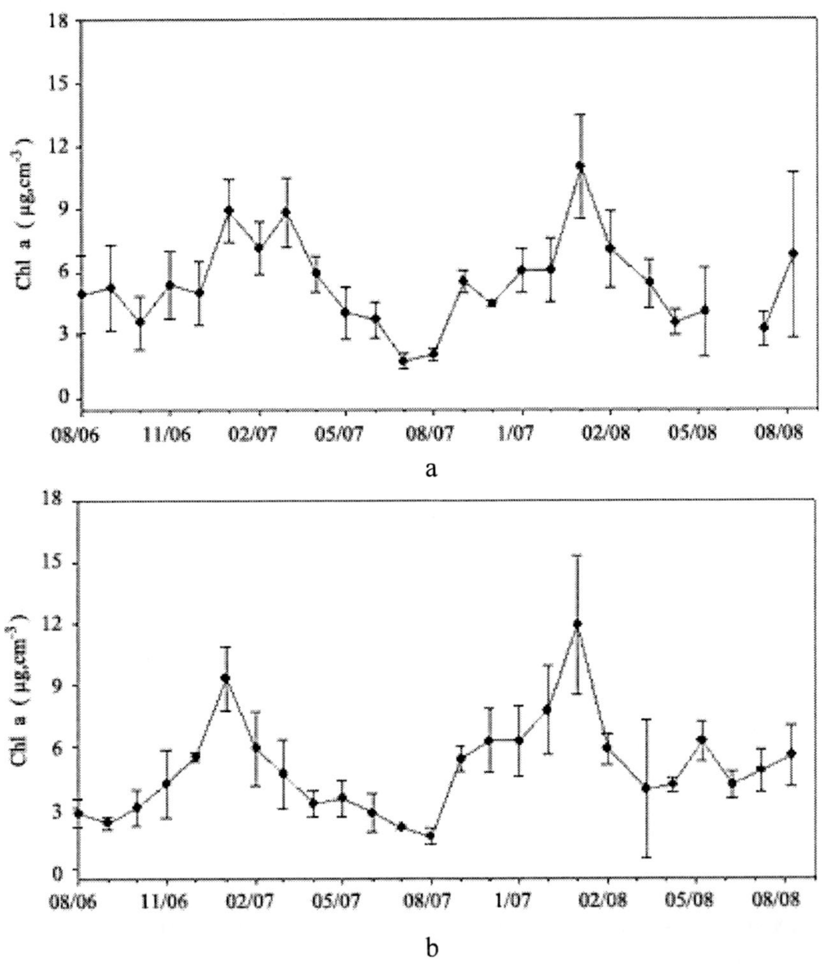

Figure 2. (Continued).

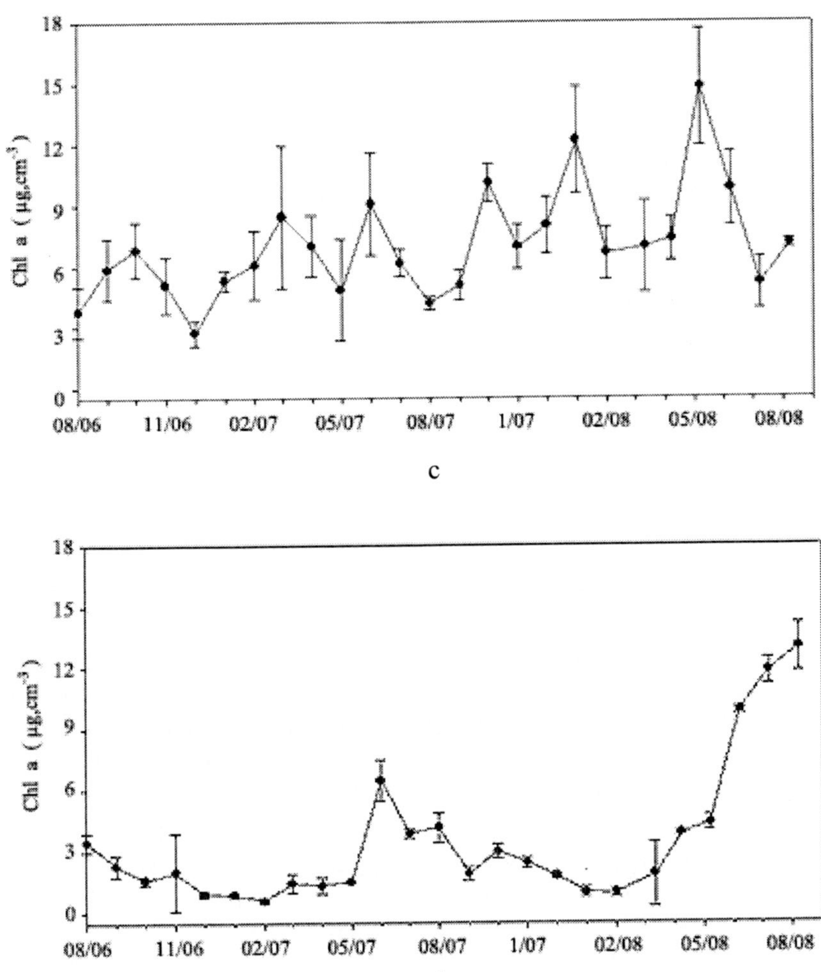

Figure 2. Seasonal variation of MPB biomass in four sampling sites (Chlorophyll *a* concentration in upper 1 cm sediment).

In this study, Microphytobenthic biomass was estimated by chlorophyll a concentration (μg cm^{-3}) considering the MPB three-dimensional distribution in sediment, which varied from 0.47 to 16.58 μg cm^{-3}. The values are in the range of those reported by Kang et al. (2007), who conducted an investigation near site C using high-performance liquid chromatography (HPLC), and comparable to those of Montani et al. (2003), who studied an estuarine sand flat of Seto Inland Sea, Japan.

Multivariate correlation analysis revealed that the biomass was positively correlated with mud and very fine sand composition, and negatively related to fine sand and medium sand composition (Table 2). The positive relationship with mud and very fine sediments (< 125 μm) corroborated previous studies on higher MPB biomass supported by muddy sediments (Underwood and Kromkamp, 1999; Riaux-Gobin and Bourgion, 2002; Perkins et al., 2003; Facca and Sfriso, 2007). In relation to other environmental factors (i.e., 4 kinds of nutrients, salinity and light intensity), their correlation with biomass was not statistically significant ($P > 0.05$). It might contribute to dramatic variation on these factors themselves in intertidal flats, as well other factors such as grazing pressure.

Vertical Distribution

Generally, the MPB biomass decreased with depth following the exponential decline models: $C_z = C_0 \times \exp(-b \times z)$. This decline pattern was consisted with previous studies (de Jonge and Colijn, 1994; Brotas and Serôdio, 1995; Montani et al., 2003). The regression curves could explain 80.4 ± 14.0 %, 84.3 ± 13.8 %, 87.4 ± 6.1 % and 73.4 ± 32.0% of the observed monthly data at sites A, B, C and D, respectively.For the average data over investigated year (August 2006-August 2007), the C_0, b, r^2 and p-value of each site was illustrated on the Figure 3.

Table 1. Multivariate correlation analyses on relationships between biomass and temperature of each site

	A	B	C	D
Chl *a*/ Temp.	-0.506[*]	-0.612[**]	NS	0.764[**]

[*], [**], significant at the 0.05 level and 0.01 level, respectively.

Table 2. Multivariate correlation analyses on relationships between biomass and environmental factors

	Mud	VFS	FS	MS	Biot	NH_4^+	PO^{3-}_4	$NO^-_2 + NO^-_3$	SiO^-_3	Light
chl*a*	0.383[**]	0.330[*]	-0.343[**]	-0.392[**]	0.106	0.010	0.064	-0.216	-0.192	0.071

[*], [**], significant at the 0.05 level and 0.01 level, respectively.

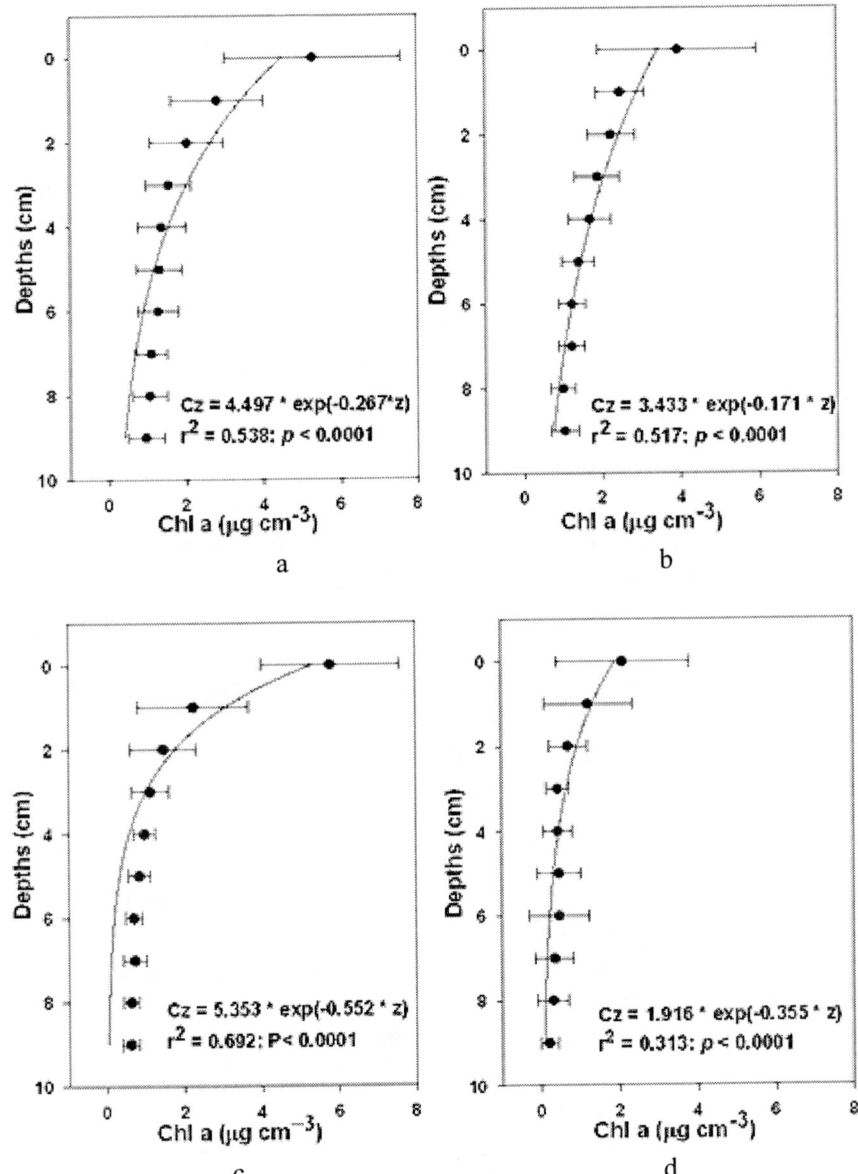

Figure 3. Regressions curves of each site on biomass with depths (C_0, C_Z are chlorophyll *a* concentration at depth 0 and z; b is curve slope, viz. the reciprocal of depth); site A, a; site B, b; site C, c; Site D,d.

Among the four study sites, site C had the largest exponential curve slope and the highest surface biomass; on the contrary, site D had the smallest slope and the lowest surface biomass (Figure 3).This difference between sandy and muddy sediments also was reported in previous studies (MacIntyre, et al. 1996; Lucas et al., 2000).

Parameter b was positively correlated to C_0, chlorophyll *a* concentration in upper 1 cm sediment and very fine sand content, and negatively correlated to fine sand content ($P < 0.05$).

The positive correlations of b with C_0 and surface chl *a* (in 1 cm) implied that high surface biomass reduced the resuspension of sediment and slowed down the burying of MPB themselves, which therefore increase the slope of the model and the stability of sediment (Miller et al., 1996; Paterson et al., 2000; Lundkvist et al., 2007).

On temporal variation in vertical distribution during a whole tidal and diurnal cycle, a certain degree of migratory rhythm was illustrated well by biomass proportion of every 0.5mm to total depth (Figure 4). Such as the migration down to a sediment depth of 2.0-3.5 mm during high tide before sunset in April 2007 and migration up to surface during low tide emersion at daytime (Figure 4).

In January 2008, the migratory rhythm in January 2008 was different except the up-migration at daytime emersion peroid. The MPB always concentrated in the surface sediment with visible thick MPB biofilms in situ, even during high tide submersion and at night time (Figure 4). Comparing two diel investigations on vertical distribution, the distribution in April 2007 was more homogeneous with average low biomass (<14 μg g^{-1}), while showed a strong stratification especially in the upper 1 mm sediment in January 2008 with highest values of 60.63 μg g^{-1} at the surface (0-0.5 mm). Without significant change of sediment composition between two diel samplings, it can be speculated that in January 2008 extremely high MPB biomass also dramatically decreased the light penetration and resuspension rates, in return kept MPB staying sediment surface.

Although the migratory rhythm in April 2007 was similar with the study of Ní Longphuirt et al. (2009) in August 2007 in the same area and most of other studies on MPB migration (Round and Palmer, 1966; Kingston, 2002; Sauer et al., 2002). As generalized by Consalvey et al. (2004), there is no universal pattern for migration, especially in sandy sediment under more complex hydrodynamic forcing and bedform morphology (Easley et al., 2005). On the other hand, our study indicated that high biomass of MPB themselves also a primary determining factor on vertical migration.

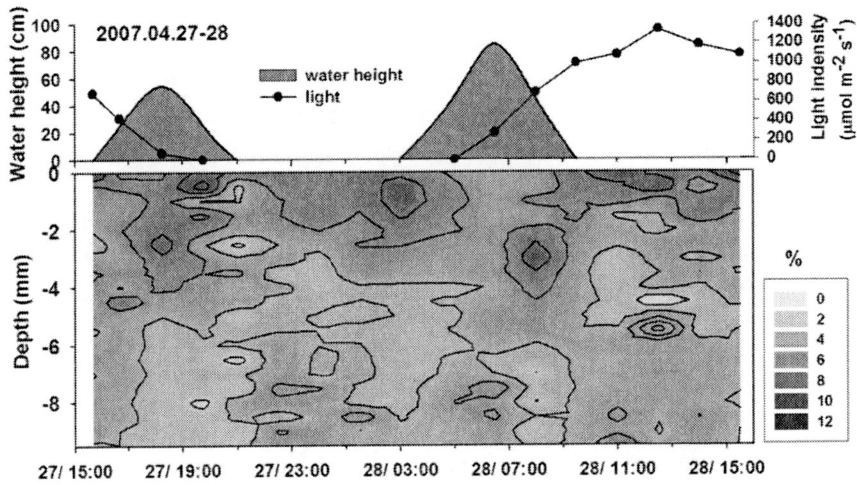

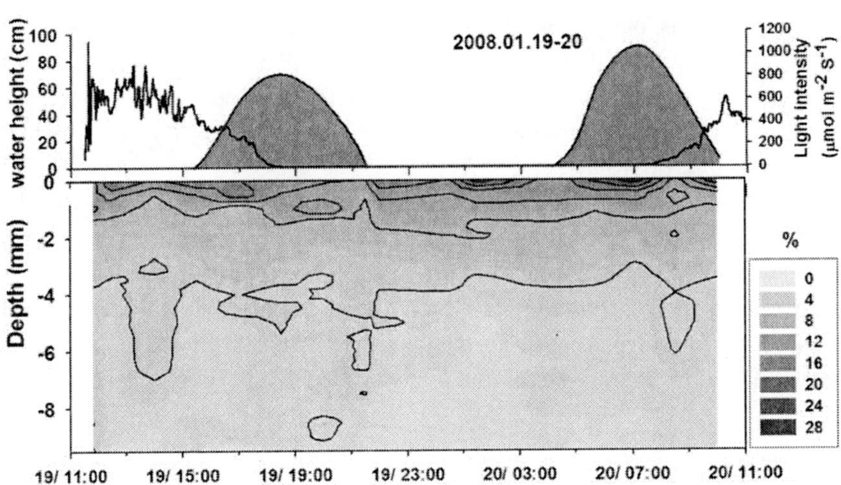

Figure 4. Temporal variation in vertical distribution along a whole tidal and diurnal cycle at muddy sand sites A. (Upper graphs is the water height over the sediment and light intensity variation during the same period. Lower contours are based on the relative percentage of chl *a* in every 0.5 mm slice to total 10 mm depths).

With dominant motile diatoms, MPB movement is determined by numerous factors, including the inducing factors such as light, tide and nutrients (Kingston, 2002; Saburova and Polikarpov, 2003; Consalvey et al., 2004; Ní Longphuirt et al., 2006), and influencing factors such as temperature, salinity and substrata type (Harper, 1969; Hay et al., 1993; Cohn and Disparti, 1994; Sauer et al., 2002), as well as physical factors that resuspend MPB during the tidal cycle. Therefore, the distribution of MPB is more complex in intertidal flats where these environmental factors fluctuate dramatically following tidal and diurnal cycles. However, defined by combination of hydrodynamic forcing and bedform morphology, the sediment type is a major abiotic factor on dynamics of MPB. Our study revealed the relationship between sediment and MPB biomass variation in 3 directions and temporal, consisting with previous studies (Oh and Koh, 1995; Cahoon et al., 1999; Mitbavkar and Anil, 2002; Stal, 2003; Facca and Sfriso, 2007; Jesus et al., 2006).

Conclusion

Our study revealed MPB community dynamics in intertidal flats as results from interactions between biotic and environmental factors. On spatial and temporal distribution, the seasonal variation of MPB showed that the biomass was positively correlated with sediment less than 125 μm, but not related to dissolved inorganic nutrients, salinity and light intensity which highly fluctuate in intertidal flats.

On vertical distribution, MPB biomass declined exponentially with depth which also related to sediment type. Diel vertical migrations presented a certain degree of rhythm following tidal and light cycles, but could be strongly weakened due to extremely high biomass at the very surface. Defined by combination of hydrodynamic forcing and bedform morphology, the sediment type is a major abiotic factor on dynamics of MPB, even could be as an indicator.

Acknowledgments

This research was financially supported by the Korean Ministry of Environment as "The Eco-technopia 21 Project" (Project No. 050010013), the Korean Ministry of Land, Transport and Maritime Affairs as "Greenhouse Gas Emission Reduction using Seaweeds". The completing of this manuscript was supported by "National Natural Science Foundation, China" (41276137), "Shandong Provincial Natural Science Foundation, China" (ZR2011CM 018).

References

Blanchard, G. F., Paterson, D. M., Stal, L. J., Richard, P., Galios, R., Heut, V., Kelly, J. A., Honeywill, C., de Brouwer, J., Dyer, K., Seguignes, M., 2000. The effect of geomorphological structures on potential biostabilisation by microphytobenthos on intertidal mudflats. *Continental Shelf Research* 20, 1243–1256.

Brotas, V., Serôdio, J., 1995. A mathematical model for the vertical distribution of chlorophyll *a* in estuarine intertidal sediments. *Netherlands Journal of Aquatic Ecology* 29, 315–321.

Brotas, V., Cabrita, T., Portugal, A., Serôdio, J., Catarino, F., 1995. Spatio-temporal distribution of the microphytobenthic biomass in intertidal flats of Tagus Estuary (Portugal). *Hydrobiologia* 300/301, 93–104.

De Brouwer, J. F. C., Bjelic, S., de Deckere, E. M. G. T., Stal, L. J., 2000. Interplay between biology and sedimentology in a mudflat (Biezelingse Ham, Westerschelde, The Netherlands). *Continental Shelf Research* 20, 1159–1177.

Christie, M., Dyer, K., Blanchard, G., Cramp, A., Mitchener, H., Paterson, D. M., 2000. Temporal and spatial distributions of moisture and organic contents across as macro-tidal mudflat. *Continental Shelf Research* 20, 1219–1241.

Cahoon, L. B., Nearhoof, J. E., Tilton, C. L., 1999. Sediment grain size effect on benthic microalgal biomass in shallow aquatic ecosystems. *Estuaries* 22, 735–741.

Cohn, S. A., Disparti, N. C., 1994. Environmental factors influencing diatom cell motility. *Journal of Phycology* 30, 818–828.

Consalvey, M., Paterson, D. M., Underwood, G. J. C., 2004. The ups and downs of life in a benthic biofilm: migration of benthic diatoms. *Diatom Research* 19,181–202.

Du, G. Y., Son, M., Yun, M., An, S., Chung, I. K., 2009. Microphytobenthic biomass and species composition in intertidal flats of the Nakdong River estuary, Korea. *Estuarine, Coastal and Shelf Science* 82, 663–672.

Facca, C., Sfriso, A., 2007. Epipelic diatom spatial and temporal distribution and relationship with the main environmental parameters in coastal waters. *Estuarine, Coastal and Shelf Science* 75, 35–49.

Easley, J. T., Hymel, S. N., Plante, C. J., 2005. Temporal patterns of benthic microalgal migration on semi-protected beach. *Estuarine, Coastal and Shelf Science* 64, 486–496.

Eleftheriou, A., McIntyre, A., 2005. *Methods for the study of Marine Benthos* (Third edition). Blackwell Science, Oxford, 418 pp.

Guarini, J. M., Blanchard, G. F., Bacher, C., Gros, P., Riera, P., Richard, P., Gouleau, D., Galois, R., Prou, J., Sauriau, P. G., 1998. Dynamics of spatial patterns of microphytobenthic biomass: inferences from geostatistical analysis of two comprehensive surveys in Marennes-Oléron Bay (France). *Marine Ecology Progress Series* 166, 131–141.

Harper, M. A., 1969. Movement and migration of diatoms on sand grains. *British Phycological Journal* 4, 97–103.

Hay, S. I., Maitland, T. C., Paterson, D. M., 1993. The speed of diatom migration through natural and artificial substrata. *Diatom Research* 8, 371–384.

Jesus, B., Mendes, C. R., Brotas, V., Paterson, D. M., 2006. Effect of sediment type on microphytobenthos vertical distribution: Modeling the productive biomass and improving ground truth measurements. *Journal of Experimental Marine Biology and Ecology* 332, 60–74.

De Jonge, V. N., Colijn, F., 1994. Dynamics of microphytobenthos biomass in the Ems estuary. *Marine Ecology Progress Series* 104, 185–196.

Kang, Ch. K., Choy, E. J., Paik, S.-K., Park, H. J., Lee, K. S., An, S. M., 2007. Contributions of primary organic matter sources to macroinvertebrate production in an intertidal salt marsh (*Scirpus triqueter*) ecosystem. *Marine Ecology Progress Series* 334, 131–143.

Kingston, M. B., 2002. Effect of subsurface nutrient supplies on the vertical migration of *Euglena proxima* (Euglenophyta). *Journal of Phycology* 38, 872–880.

Kromkamp, J. C., De Brouwer, J. F. C., Blanchard, G. F., Forster, R. M., Creach, V., 2006. *Functioning of microphytobenthos in estuaries*. Royal Netherlands Academy of Arts and Sciences, Amsterdam, 262 pp.

Lorenzen, C. J., 1967. Determination of chlorophyll and pheo-pigments: Spectrophotometric equations. *Limnology and Oceanography* 12, 343–346.

Lucas, C. H., Widdows, J., Brinsley, M. D., Salkeld, P. N., Herman, P. M. J., 2000. Benthic-pelagic exchange of microalgae at a tidal flat. 1. Pigment analysis. *Marine Ecology Progress Series* 196, 59–73.

Lundkvist, M., Grue, M., Firend, P. L., Flindt, M. R., 2007. The relative contribution of physical and microbiological factors to cohesive sediment stability. *Continental Shelf Research* 27, 1143–1152.

MacIntyre, H. L., Geider, R. J., Miller, D. C., 1996. Microphytobenthos: The ecological role of the "secret garden" of unvegetated, shallow-water marine habitats. I. Distribution, abundance and primary production. *Estuaries* 19, 186–201.

Méléder, V., Rincé, Y., Barillé, L., Gaudin, P., Rosa, P., 2007. Spatio-temporal changes in microphytobenthos assemblages in a macrotidal flat (Bourgeneuf Bay, France). *Journal of Phycology* 43, 1177–1190.

Miller, D. C., Geider, R. J., MacIntyre, H. L., 1996. Microphytobenthos: The ecological role of the "secret garden" of unvegetated, shallow-water marine habitats. II. Role in sediment stability and shallow-water food webs. *Estuaries* 19, 202–212.

Mitbavkar, S., Anil, A. C., 2002. Diatoms of the micropytobenthic community: population structure in a tropical intertidal sand flat. *Marine Biology* 140, 41–57.

Mitbavkar, S., Anil, A. C., 2004. Vertical migratory rhythms of benthic diatoms in a tropical intertidal sand flat: influence of irradiance and tides. *Marine Biology* 145, 9–20.

Montani, S., Magni, P., Abe, N., 2003. Seasonal and interannual patterns of intertidal microphytobenthos in combination with laboratory and areal production estimates. *Marine Ecology Progress Series* 249, 79–91.

Ní Longphuirt, S., Leynaert, A., Guarini, J.-M., Chauvaud, L., Claquin, P., Herlory, O., Amice, E., Huonnic, P., Ragueneau, O., 2006. Discovery of microphytobenthos migration in the subtidal zone. *Marine Ecology Progress Series* 328, 143–154.

Ní Longphuirt, S., Lim, J. H., Leynaert, A., Claquin, P., Choy, E. J., Kang, C. K., An, S., 2009. Dissolved inorganic nitrogen uptake by intertidal microphytobenthos: Nutrient concentrations, light availability and migration. *Marine Ecology Progress Series* 379, 33–44.

Oh, S. H., Koh, C. H., 1995. Distribution of diatoms in the surficial sediments of the Mangyung-Dongjin tidal flat, west coast of Korea (Eastern Yellow Sea). *Marine Biology* 122, 487–496.

Parsons, T. R., Maita, Y., Lalli, C. M., 1984. *A manual of chemical and biological methods for seawater analysis.* Pergamon Press, Oxford, 173 pp.

Paterson, D. M., Tolhurst, T. J., Kelly, J. A., Honeywill, C., de Deckere, E. M. G. T., Huet, V., Shayler, S. A., Black, K. S., de Brouwer, J., Davidson, I., 2000. Variations in sediment properties, Skeffling mudflats, Humber Estuary, UK. *Continental Shelf Research* 20, 1373–1396.

Perissinotto, R., Nozais, C., Kibirige, I., 2002. Spatio-temporal dynamics of phytoplankton and microphytobenthos in a South African temporarily-open estuary. *Estuarine, Coastal and Shelf Science* 55, 47–58.

Perkins, R. G., Honeywill, C., Consalvey, M., Austin, H. A., Tolhurst, T. J., Paterson, D. M., 2003. Changes in microphytobenthic chlorophyll *a* and EPS resulting from sediment compaction due to de-watering: opposing patterns in concentration and content. *Continental Shelf Research* 23, 575–586.

Riaux-Gobin, C., Bourgoin, P., 2002. Microphytobenthos biomass at Kerguelen's Land (Subantarctic Indian Ocean): repartition and variability during austral summers. *Journal of Marine Systems* 32, 295–306.

Round, F. E., Palmer, J. D., 1966. Persistent, vertical-migration rhythms in benthic microflora. II. Field and laboratory studies on diatoms from the banks of the river Avon. *Journal of the Marine Associationof the United Kingdom.* 46, 191–241.

Saburova, M. A., Polikarpov, I. G., 2003. Diatom activity within soft sediments: behavioural and physiological processes. *Marine Ecology Progress Series* 251, 115–126.

Sauer, J., Wenderoth, K., Maier, U. G., Rhiel, E., 2002. Effects of salinity, light and time on the vertical migration of diatom assemblages. *Diatom Research* 17, 189–203.

Serôdio, J., Catarino, F., 2000. Modelling the primary productivity of intertidal microphytobenthos: time scales of variability and effects of migratory rhythms. *Marine Ecology Progress Series* 192, 13–30.

Stal, L. J., 2003. Microphytobenthos, their extracellular polymeric substances, and the morphogenesis of intertidal sediments. *Geomicrobiology Journal* 20, 463–478.

Thornton, D. C. O., Dong, L. F., Underwood, G. J. C., Nedwell, D. B., 2002. Factors affecting microphytobenthic biomass, species composition and production in the Colne Estuary (UK). *Aquatic Microbial Ecology* 27, 285–300.

Underwood, G. J. C., 1994. Seasonal and spatial variation in epipelic diatom assemblages in the Severn Estuary. *Diatom Research* 9, 451–472.

Underwood, G. J. C., Kromkamp, J., 1999. Primary production by phytoplankton and microphytobenthos in estuaries. *Advances in Ecological Research* 29, 93–153.

Underwood, G. J. C., Philips, J., Saunders, K., 1998. Distribution of estuarine benthic diatom species along salinity and nutrient gradients. *European Journal of Phycology* 33,173–183.

Index

D

E

F